Site engineering

ROY W. MURPHY, MCIOB

Construction Press

LONDON AND NEW YORK

Construction Press
An imprint of:
Longman Group Limited
Longman House, Burnt Mill, Harlow
Essex CM20 2JE, England
Associated companies throughout the world

*Published in the United States of America
by Longman Inc., New York*

© Construction Press, 1983

First published 1983

British Library Cataloguing in Publication Data
Murphy, Roy W.
  Site engineering. – (Site practice series)
  1. Building sites – Management
  I. Title   II. Series
  624.068   TH438

  ISBN 0-582-40606-4

Library of Congress Cataloging in Publication Data
Murphy, Roy W., 1951–
  Site engineering.

  (Site practice series)
  Bibliography: p.
  Includes index.
  1. Building sites. I. Title. II. Series.
TH375.M87   690'.11   82-1445
ISBN 0-582-40606-4   AACR2

Printed in Hong Kong
by Astros Printing Limited

Site engineering

SITE PRACTICE SERIES

General editors: Harold Lansdell, FCIOB, FCIArb, and
  Win Lansdell, BA

Site safety – *Jim Laney*
Site engineering – *Roy Murphy*

Books to be published in the Series
Making and placing concrete – *Edwin Martin Baker*
Timber frame housing – *Jim Burchall*
Security on site – *Leonard Earnshaw*
Site carpentry and joinery – *Keith Farmer*
Site relations – *Tom Gallagher*
Careers in the building industry – *Chris and Lynne March*
Fixings, fastenings and adhesives – *Paul Marsh*
Glazing – *Stanley Thompson*
Steel reinforcement – *Tony Trevorrow*
Exercises in brickwork and blockwork – *Arthur Webster*

Contents

Preface

The site engineer is considered by the author to be a key member of the construction site team. He is not merely the provider of lines and levels required to establish the setting out of the works but should organize and control aspects of the construction, particularly up to finish floor level. Thus the site engineer could be seen as performing a role similar to that of the site agent, but for works below floor level.

The book covers most of the general responsibilities of the site engineer and practical advice is given throughout to assist him in his task.

It is intended that the reader, whether he be a trainee or experienced engineer will find the book useful, in both approach and detail, for all aspects of setting out.

Modern construction techniques, involving the use of system buildings and prefabricated components, mean that accurate setting out is more important than ever and the engineer must practise the procedures necessary to achieve this degree of accuracy. He must also undertake the routine checking methods explained throughout the text.

The formulae and techniques explained are those considered adequate to achieve the required tolerances, and it is essential that the reader familiarizes himself with the operation of each.

The early chapters contain information on the surveying equipment available and frequently needed, and relate to its operation prior to considering its use on site.

Those chapters explaining practical methods of setting out and recording of details are arranged to conform to the usual order of operations on sites of medium size.

The setting out procedures considered are suitable for the average size and type of project, such as, schools, hospitals, factories and housing sites, which are frequently built for local authorities and development corporations. It is not intended that the book should cover fully the techniques required on large civil engineering or land surveying operations although many of the same principles may still apply, albeit on a larger scale. Similarly, smaller contracts are not considered as the engineer is not generally employed on such projects; the site foreman will construct the works to an acceptable accuracy by using more basic setting out procedures.

The use of computers by architects to design sites, is becoming increasingly popular and setting out details are often presented in a format that requires further calculations on site. The use of programmable pocket calculators is referred to throughout the text because it is hoped that the engineer will endeavour to use them for many aspects of his work. Procedures for formulating programs are explained and examples are given of the types of program that can be compiled. The advantages gained by using these calculators are numerous and the programming technique required gives an insight into 'BASIC' computer language. This may be of value to the engineer employed by a progressive company which uses computers.

Acknowledgements

My thanks go to all those who have contributed, assisted and given their consent to the use of their material within the text. Photographs and details of Wild surveying instruments (Figs. 1.4, 1.5, 1.7, 1.8, 1.9 (b) and (c), 2.9, 2.10 and 2.11) are reproduced by kind permission of Wild Heerbrugg (United Kingdom) Limited, Revenge Road, Lordswood, Chatham, Kent ME5 8TE. Details and photographs of the Texas TI57 programmable calculator (Fig. 1.3) are adapted and reproduced by permission of Texas Instruments Limited, Manton Lane, Bedford MK41 7PU. J. Uren and W. F. Price of Portsmouth Polytechnic (Department of Civil Engineering), gave permission to reproduce an adapted version of Table 4.2 'Angle booking sheet', from their book *Surveying for engineers* (Fig. 3.2). CIRIA (Construction Industry Research and Information Association) gave permission to reproduce adapted versions of diagrams (Figs. 6.14 and 6.15), Figs. 25 and 26, tables and details of scale factor calculations, from their publication *A manual of setting out procedures*. Material from BS 5606:1978 is adapted and reproduced by permission of the British Standards Institution, 2 Park Street, London W1A 2BS, from whom complete copies can be obtained. Material from BRE Digest 202 (June 1977) is adapted and reproduced by permission of the Building Research Station, Garston, Watford WD2 7JR from whom complete copies can be obtained. My thanks to Edward Arnold (Publishers) Limited for adapted versions of level booking procedure (Fig. 3.1) taken from their publication *Engineering surveying problems and solutions* by F. A. Shepherd. AGA Geotronics, 6 The Quay, St. Ives, Cambridgeshire, for kind permission to reproduce photographs (Figs. 1.9 (a) and 1.10) of their Geodimeter Model 120 and Geoplane 300 surveying instruments. Clarksons Ltd, 1 Brixton Hill Place, London SW2 1HL, for permission to reproduce (Fig. 2.8) diagram showing the layout of controls on a 'Sokkisha TM20E/ES theodolite, distributed by Clarksons in the UK. Finally I extend my appreciation to those who during my career have given their time and guidance in my own training.

1

Surveying equipment

During his work the site engineer will undertake the measurement of angles, distance and height. Using measured lengths and angles, together with evaluated levels he will establish the required location of points in the field. Equipment has been devised to enable him to perform these tasks; it has been employed throughout the history of surveying and forms a tool kit for the engineer.

The site engineer working with a progressive and modern contractor should aim to have the basic equipment available for use as listed in (A) below together with a selection of the surveying equipment listed in (B).

(A) Basic equipment

(a) steel tape (open frame type if possible) 30–50 m in length
(b) steel tape 3 m in length (pocket type)
(c) ranging poles (or rods) 2–2.5 m in height
(d) spirit level
(e) plumb-bob
(f) string lines
(g) 14 lb sledge hammer
(h) claw hammer
(i) crayons or thick felt-tip pens
(j) programmable pocket calculator
(k) field books and level books
(l) paint of various colours
(m) planboards, clipboard, and polythene pockets
(n) kitbag.

(B) Surveying equipment

(o) level
(p) theodolite
(q) optical plummets and autoplumbs
(r) E.D.M. (electromagnetic distance measuring equipment)
(s) laser line and level equipment.

Obviously, depending on the size of the company and the type of work undertaken, all or part of the equipment listed may be provided. However, to achieve satisfactory results, all items except (q) (r) (s), should be purchased. Items (q) (r) (s) should be supplied as and when the setting out dictates their use.

Before looking at the use of the main items of surveying equipment e.g. tapes, levels, theodolites, E.D.M., and calculators, in detail in Chapter 2, we will consider equipment in general.

Evaluation of basic equipment

(a) Steel tape (30 m – 50 m length)

The choice of steel tape manufacturers is large, and as these items are subject to rough use it is advisable to select a type that is not only accurate but capable of withstanding the rigours of site work. Given adequate attention, regular cleaning, and an occasional light coating of oil, most reputable manufacturers' tapes prove durable.

However, many users find the steel band type with an open winding frame and folding handle preferable – it is less prone to damage when winding in, especially if compared to the cased tapes which can become congealed with mud after only a morning's use. When working in wet and muddy conditions, the tape should be wound in whilst its band is held between a rag, thus preventing mud from entering the reel.

Most steel tapes are 10–12 mm wide and have either etched steel or enamelled markings in 1 or 5 mm increments. Tapes should be purchased in the 30–50 m length size.

Linen and plastic tapes are not recommended where accurate setting out dimensions are needed as they are prone to stretching and other inaccuracies. Details of taping procedures are given in Chapter 2.

(b) Steel tapes (3 m length)

The engineer will invariably find the need for a 3 m long extending steel tape such as a 'Stanley Powerlock'. When measuring short distances, making travellers and profiles, or adjusting small measured discrepancies, such tapes are very useful as they are easier to handle than the larger sizes.

(c) Ranging poles

These are wooden or metal poles made in various lengths although usually purchased in the 2–2.5 m size, and painted alternately in 500 mm increments of red and white, or red, black and white. The lower tip of the pole is pointed and metal clad to penetrate the earth.

The poles are used for sighting a straight line between two established points and must be erected vertically. The engineer stands behind one pole and signals to an assistant who inserts other ranging rods to align with a pole located at the opposite end (see Fig. 1.1). It is more accurate for the assistant to start locating

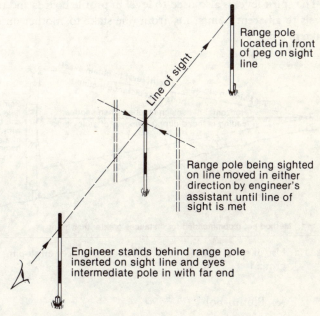

Range pole located in front of peg on sight line

Range pole being sighted on line moved in either direction by engineer's assistant until line of sight is met

Engineer stands behind range pole inserted on sight line and eyes intermediate pole in with far end

Line of sight

Fig. 1.1 Sighting of range poles

3

intermediate poles from the far end working towards the engineer. Always sight to the lower part of the pole as this will reduce errors if the rod is not being held truly plumb.

Ranging rods can be used as markers for pegs located in tall grass or thick undergrowth, and as measuring rods for rough dimensions. For hard ground ranging pole support legs can be purchased – these are of light robust construction and are in the form of a tripod.

(d) Spirit level

A 1 m or similar length spirit level should be used to vertically plumb between points e.g. taping short distances on sloping ground. For such instances it is unadvisable to locate pegs at more than 10 m intervals. When taping on slopes, the engineer should drive in his peg at the rough location and accurately mark the nail point by asking his assistant to hold a spirit level (braced with the shaft of his sledge hammer or timber) on top of the peg. The engineer then swings the tape up and down in an arc to establish the required length (see Fig. 1.2).

The spirit level is also used to level in profile boards and transfer levels to adjacent points, i.e. from one stake to another on double stake profiles.

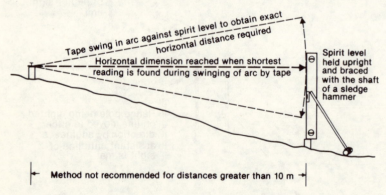

Fig. 1.2 Use of spirit level for taping of short distances on sloping ground

(e) Plumb-bob

The plumb-bob is perhaps used less today than it was some years ago but still forms part of the kitbag. Mainly used for setting up

4

instruments over a fixed point such as a nail head, otherwise for locating a point some distance below an overhead object, or as a weight to establish a vertical line down a wall face.

(f) String lines

Good quality nylon string lines should be purchased as they are durable in the thin section required (such strings are often sold as bricklayers lines). Their uses are numerous, e.g. stringing between pegs to denote foundation dig and check lines, to produce top level line in road kerb construction and to mark the brickwork lines etc.

(g) Sledge hammer

A 14 lb sledge hammer is preferred to drive pegs and stakes into most ground strata.

(h) Claw hammer

Use the claw hammer when affixing profile boards and locating nails in pegs etc. Dismantling of profiles will also be required and all old nails should be removed.

(i) Crayons and thick felt-tip pens

Heavy duty wax crayons are ideal for marking nail positions and referencing locations on concrete and brick etc.

Thick felt-tip pens can be used to write on timber to denote the location of pegs, chainage position or setting out points. Profiles may have their level or height above reduce excavation written on to the cross board.

Both crayons and felt-tip pens are fairly durable, although for more permanent marked positions, paint should be used. Road marking paint or blackboard paint are suitable materials.

(j) Programmable pocket calculator

The engineer may at first consider a standard or scientific calculator adequate. However, after tackling the calculations required for co-ordinate and road works, he will soon realise that a programmable calculator is a much better prospect. For only a relatively small extra outlay, basic programmable calculators having up to 50 or more program steps can be purchased.

The Texas TI57 calculator (see Fig. 1.3) is an example of an inexpensive machine capable of performing many of the calcu-

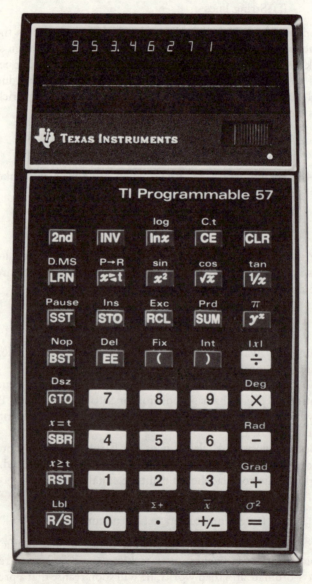

Fig. 1.3 Texas TI57 programmable pocket calculator

lations the engineer will encounter in general construction works. Obviously the choice of such calculators is vast and being constantly updated, although the examples of programs shown in later chapters are for use with the Texas TI57.

Whichever calculator is purchased the user must become fully conversant with its operation, memory, functions, modes and programming techniques. The capability to devise one's own programs gives major advantages in time saving and accuracy, and makes checking easier.

Companies engaged in large survey and construction projects may even have (or find it useful to purchase) their own computer, which can be put to use on evaluation and processing of engineering calculations.

(k) Field books and level books

The keeping of accurate and neat field books is of vital importance. Printed field books for levelling and surveying, with stout pages and durable covers are readily available. Books for either 'rise and fall' or 'collimation' method level logging may be obtained, depending on the method of calculation the user prefers. Some instrument manufacturers also publish their own brand of field book.

Calculations should be carried out in a logical manner and set out in a neat format, this will reduce the possibility of errors and facilitate ease of checking. If an engineer moves contracts and hands over his current project to a colleague, the library of completed field and level books should be passed on.

(l) Coloured paint

To identify pegs, profiles, service ducts and numerous other items, various colours of paint should be used. Colour coding for pegs marking various sections of the works assists recognition and allows ease of reference for the foreman and operatives who use them. The colour coding used on a housing project could be as follows

    (a) White for house pegs
    (b) blue for storm drainage trench centre lines
    (c) red for foul drainage trench centre lines
    (d) yellow for roads
    (e) green for services trench centre lines.

Similar colour coding for road service crossing ducts should also be devised, the kerbs being marked as the services are laid.

(m) Planboards, clipboard and polythene pockets

The method of transporting the engineer's calculated information or sketched diagrams from office to site needs careful consideration. It is pointless to spend time carefully preparing neat sketches and compiling detailed traverse sheets, only to see them spoilt when taken on site by rain, mud, creasing or similar hazards. To safeguard his paperwork and avoid such problems, the engineer can adopt the following methods of protection.

Planboards

Planboards can be cut from either hardboard or thin plywood to facilitate transportation. They can be made to various sizes to suit the details requiring display. The plan should be Sellotaped to the board and covered by clear polythene sheeting of adequate strength and this in turn should be affixed to the board, by tape or staples.

Clipboards and polythene pockets

Clipboards used to keep together the smaller detail and calculation sheets collected by the engineer may be transported around site, a different page being displayed to cover each section of the day's tasks. The information, if written or sketched on A4 size paper and forms, can be contained within a clear polythene pocket such as those used for protection of material held in ringbinders etc.

(n) Kitbag

A stout kitbag with a shoulder strap is a useful purchase. Ex-military type ration bags often with light metal baseplates and ribbing are cheaply available from army and navy stores. These make ideal holdalls for the field book, nails, marker pens, string line, claw hammer and other small items used daily by the engineer.

Evaluation of surveying equipment

(o) The level

There are various types of levelling instrument which may be provided by the engineer's employers. All the many brands have different methods of operation, although the basic procedures are similar. The method of setting up will be the same and is explained in Chapter 2. The engineer must become fully conversant with the operation of his particular type of level. Some instruments, especially the older types, show an inverted image, whilst most modern types with bloomed lenses give an erect image. The use of different instruments with different modes of operation is not advised because it can lead to error. An engineer should use one instrument only except when his own level is undergoing maintenance.

Level types

The Cowley level The Cowley level is used for short distance levelling (i.e. less than 60 m) during drainage and small scale construction work. The instrument when erected is operated by viewing onto mirrors through an eyepiece mounted on top of the instrument. Sets of mirrors within the level are fixed to the case, whilst another is in suspension so as to render the instrument level irrespective of its angle (up to a limit). A horizontal bar target is viewed through the mirror image which splits the bar into two halves, the target bar is then moved up and down until both its sides coincide when seen through the level.

The dumpy level The dumpy level was the standard levelling instrument used for levelling over the 60 m range of a Cowley level. However, it has been repeatedly modified and is now little used in its basic form, although some manufacturers still retail instruments termed dumpy levels. The original type was basically a telescope mounted upon a tribrach and, once levelled by the use of a graduated bubble, revolved about a vertical axis in a horizontal plane needing no other adjustment until relocated at a new set up position.

The dumpy level has been superseded by the tilting, quickset, precise and automatic levels. An example of a modern type dumpy level is shown in Fig. 1.4.

The tilting level The tilting level updated the dumpy level by giving the facility of collimation line adjustment in relation to the vertical axis. The instrument still has a three screw tribrach base which incorporates a centre circle bubble for initial levelling purposes. Accurate levelling is controlled by means of the tilting screw which adjusts the pivoted telescope and sensitive graduated

Fig. 1.4 Wild NK01 Modern type dumpy level

bubble tube. The increased sensitivity of the level bubble gives more accuracy in readings taken on site.

The quickset level The quickset level is an adaption of the tilting level whereby the tribrach is replaced by a ball and cup connecting mount. The instrument is roughly levelled by the use of a circular centre bubble and is then, as in the case of tilting levels, more finely adjusted using a graduated screw to move a sensitive tube bubble. Many quickset levels incorporate a circular scale on their base calibrated to 360° and used for rough angular measurement.

   The precise level Precise levels are more accurate than the quickset and dumpy level, especially when used over longer distances. The levels have either a three screw tribrach or ball and cup base, together with a circular centre bubble for the provisional levelling of the instrument. Nearly all types use the system of optical coincidence level bubbles for final accurate levelling of the instrument (see Fig. 2.2) often visible through the lens, thus eliminating the need to constantly remove one's view from the eyepiece when checking if the instrument has remained level. Most precise levels have greater magnification than the dumpy and quickset types.

   The automatic level Many models and makes of automatic level are now produced (see Fig. 1.5). Most are constructed to a high degree of accuracy, and operate by means of self levelling glass prisms within the casing of the instrument. They offer the advantages of fast operation and a reduction in error by eliminating the need to re-level the instrument each time the telescope is moved to sight the staff. Once the instrument has been basically levelled by means of a circular centre bubble, it will be self levelling until transferred from its erected location.
   Some models incorporate a horizontal graduated circle visible through a glass dome, which may be used for measurement of angles to a limited extent.

   The levelling staff
Various types of staff are available for use with levelling equipment. Differing materials, i.e. wood or metal with plastic or painted faced graduated scales (now almost always in metric) are used. Most modern staffs are now made in light alloy metal, although wooden staffs are still available.
   Staff graduations are in metres, decimeters, centimetres and if desired, half centimetres. Many staffs have a centre circle bubble for plumbing purposes. A staff divided into centimetre increments is the normal requirement, it can be split by eye to 5 mm or even less at close range, with a fair degree of accuracy.
   The R.U. (British Standard) staff is now the commonly accepted pattern, although others are available. The R.U. pattern

Fig. 1.5 Wild NA2 Universal automatic level

incorporates the 'E' symbol to denote graduations over the first 50 mm from each 100 mm increment (see Fig. 1.6). Some manufacturers provided replacement stick-on faces for staffs on which markings have become worn.

(p) The theodolite

The majority of contractors find it necessary to purchase a theodolite for measurement of vertical and horizontal angles. Theodolites can be purchased to suit the degree of accuracy required by the contractor, i.e. 1-second, 10-second or 20-second types. It is important to buy a theodolite which is accurate within the tolerances required. Theodolites incorporate stadia lines visible

Fig. 1.6 Section of R.U. (British Standard) levelling staff

through the telescope, these are provided for levelling and tacheometry purposes.

As with the level, many different makes and models are available and it is only with use that the engineer will decide which type best suits his purpose. Most leading manufacturers produce ranges from the very basic to extremely accurate models. The type purchased will depend on its intended use and cost. For small scale survey and construction projects, a 20-second theodolite may be perfectly adequate, whereas for large surveys and precise setting out, a 1-second instrument is essential.

General description of the theodolite

Modern theodolites are constructed on the same principles as their predecessor the Vernier theodolite, although the scale reading, levelling and general performance has been improved.

A telescope is mounted on a horizontal axis within a trunnion, which is supported from the alidade of the instrument by means of standards. The alidade is pivoted on a tribrach, connected to a trivet stage by three levelling footscrews. Located on the alidade is the levelling bubble for the plate. In many modern theodolites the

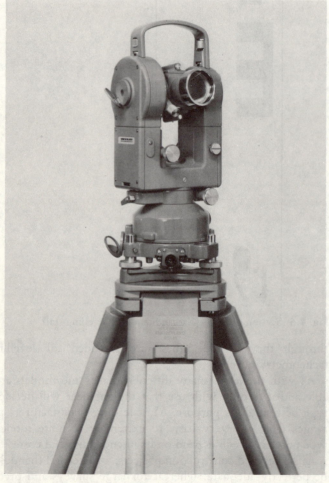

Fig. 1.7 Wild T2 theodolite

vertical levelling bubble has been replaced by an automatic index, which automatically sets the instrument for reading of vertical angles when the theodolite has been levelled by the plate level.

The various slow motion and fixing clamps are located in different places on different models, some have sliding base clamps, others have a detachable tribrach. On the most recent types of instrument the vertical and horizontal scales are photographically etched on glass, giving more accuracy than the older engraved metal scale circles. Many theodolites house an optical plummet within the alidade of the instrument allowing more accurate setting up.

Figure 1.7 shows a photograph of the Wild T2 theodolite. Before using any modern theodolite, the engineer must fully aquaint himself with its operation by reference to the manufacturer's handbook.

Figure 2.8 denotes the layout of controls on a typical theodolite.

The 1-second theodolite

The 1-second theodolite has very accurate scales which can be read to one second of a degree. Because of the restricted tolerance of these instruments they are costly to manufacture and expensive to buy. For accurate work a 1-second theodolite is essential. It is also ideal for use with electronic distance-measuring equipment. Most major instrument manufacturers, produce a one second theodolite and therefore a good selection is available for purchase.

The 20-second theodolite

Twenty-second theodolites are produced to fulfil the needs of those who may find such accuracies quite sufficient. Builders who require a theodolite merely to turn right angles, or to do the less accurate work between more accurately located stations, often find the 20-second instrument satisfactory.

The scales on many of the instruments can, with care, be visually split to an accuracy of less than 10 seconds, although such readings can only be estimated.

(q) Optical plummets and autoplumbs

Optical plummets or autoplumbs are, manufactured by many instrument makers. Some are capable of viewing both above and below. The instruments are used to accurately establish points overhead or downward, which would be difficult or impossible to locate using conventional spirit levels or plumb-bobs (see Fig. 1.8). Optical plummets are also as stated in section (p) on page 15, frequently incorporated as standard in modern theodolites and tribrachs.

Most autoplumbs are self levelling, once the user has roughly levelled the instrument by means of a centre circle bubble, and automatically establish a vertical position. Such instruments are mainly used to ensure accurate plumbing of tall structures. However, in recent years the laser beam system has become increasing popular.

Fig. 1.8 Wild ZL automatic zenith plummet

(r) E.D.M. (electromagnetic distance measurement)

Modern methods used by architects and engineers to present information in plan and book form (e.g. computer print outs and co-ordinated locations for reference and setting out points) dictates that contractors must become more technically orientated when undertaking projects presented in such a manner.

The engineer can increase the speed and accuracy with which he calculates from plans by using a programmable calculator. He must maintain this accuracy and speed when locating his points in the field and in order to do so more and more contractors are purchasing E.D.M. equipment. Previously these instruments were used mainly by surveyors covering large areas. Whether a contractor can justify investing in such a machine depends on size and type of contracts encountered: the instruments and the equipment required for their operation are very expensive. A wide range of equipment is now available: it is important that the instrument eventually purchased is the most suitable for the contractor's needs.

The prime considerations when purchasing E.D.M. equipment are

(a) accuracy
(b) reliability
(c) cost.

Measurement is by means of carrier waves, frequently the light emitting diode (infra-red) system is used with glass prism reflectors. Some E.D.M. units are adaptations for basic theodolites, but others are complete instruments sometimes called total stations or tachymats. Most run on rechargeable portable electric power packs. Theodolite adapted models are useful because when measuring facilities are not required the theodolite can still be used. However, it is important that the E.D.M. equipment matches up with the theodolite available or accuracy will be impaired.

As a guide, the following instrument performance should be aimed for when considering E.D.M. purchase for general construction use:

1 accuracy $\pm$ 5 mm + 5 p.p.m. (parts per million)
2 range – to suit desired use, but preferably at least 500 m
3 size and weight – light, compact and easy to carry
4 ease of set up – some models are difficult
5 time required for warm up and battery power life
6 instrument should incorporate
   (a) auto signal balancing
   (b) atmospheric control switch

   (c) auto digital read out

   (d) measurement quality indication

7 tracking facilities

8 good quality telescope and E.D.M. lens

9 compact and lightweight batteries with battery state indicator

10 facility for reduction of slope to plan length is an advantage

11 adaptability to other brands of equipment

12 effect on measurements if beam interrupted

Fig. 1.9 (a) AGA Geodimeter 120 mounted on Wild theodolite

13 servicing facilities within the country of use

14 frequency socket.

Examples of available E.D.M. equipment are shown in Fig. 1.9(a), (b), (c). For those interested in learning about the construction and technical background of electromagnetic distance measurement, *Electromagnetic distance measurement* (Burnside 1971) is recommended.

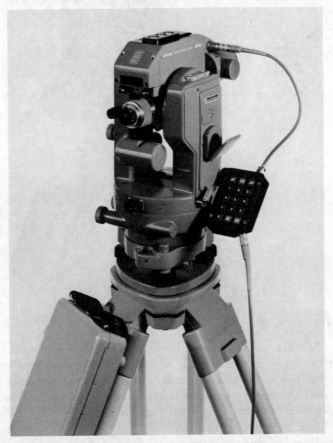

Fig 1.9   (b) Wild DI₄ Distomat

(s) Laser line and level equipment

Recent years have seen the development of the laser beam and its use within the construction industry. Laser instruments are now available to give horizontal, vertical and gradient alignment.

Fig 1.9 (c) Wild Tachymat TC₁

Most instruments operate by projecting a visible beam of light from a rotating laser head, which can be set at various scanning speeds or stopped completely at any angle. Instruments can be fixed on tripods or similar stable bases – many are designed to be self levelling after being set at a known datum.

Sensor staffs are available for level reading because the beam will be less distinct in strong sunlight. These incorporate the use of a laser detector which is automatically or manually adjusted to detect and position the laser beam in relation to the staff.

Fixed beam lasers are available for drain laying, these are capable of adjustment to suit the angle (or gradient) of the drain. The instrument is set up on the manhole base at one end (usually lowest) then directed to coincide with the line of the drain and ad-

Fig. 1.10   AGA Geoplane 300 laser

justed to suit the gradient. A target disc standing within the pipe being laid is adjusted until the visible beam hits the centre of the target.

Equipment is also available for attachment to drainage machines, graders and dozers, these allow dig levels to be controlled automatically by using laser detectors.

Lasers usually operate from a 12 V electric battery which adds to the bulk of the equipment. Due to the visible light source, lasers are particularly suitable in tunnel and minework, and laser theodolites have also proved useful in this area. Figure 1.10 shows an example of laser equipment.

2

The use of surveying instruments

In Chapter 1 we examined the types of surveying instruments available and those most frequently used by site engineers. Obviously different makes and models of instrument will vary slightly in their method of operation: in this chapter we discuss the basic principles.

Care of site instruments

All equipment, levels, tapes, theodolites, E.D.M., lasers, etc. is manufactured to precise tolerances, and whilst it may be designed to withstand rugged use on the construction site it will not stand up to abuse and neglect. Even with careful use instruments need regular cleaning and maintenance, and levels, theodolites and E.D.M. equipment require frequent checking and periodic servicing by the manufacturers.

The following sections will detail, for each instrument, those points which require particular consideration.

Measuring tapes

The tape is probably the engineer's most abused piece of equipment. The laborious task of cleaning will not be relished by the user but, such an important part of the engineer's kitbag must be properly maintained.

As suggested in Chapter 1 the tape should be cleaned when winding in and, when working in muddy conditions, each time it is used. A more thorough cleaning monthly, weekly or, during the winter, daily is desirable. Fully extend the tape and clean it with a damp cloth, thoroughly dry the band and wind in through a lightly oiled rag.

A broken or kinked tape can be repaired using a tape repair kit such as that available from Rabone Chesterman Ltd. These kits contain an assortment of metal strips of various lengths and widths, also pop-rivets and a fixing tool. The repair can be made quickly and easily and will last until a replacement is obtained. Care should be taken when repairing tapes to ensure that no loss of accuracy occurs. To ensure accuracy mark a distance of exactly one metre on a solid and level surface then compare the repaired section of the tape with this. Many manufactures produce spare tape bands and end hooks to replace bands that have become too worn to be accurately read and used. Remember that a tape with a kink or a number of bad snags in it may be inaccurate and should be repaired or replaced before being reused. Regularly check the tape for stretch over a known length of solid base.

Levelling instruments

Most levels whether dumpy, quickset or automatic, are used daily on construction sites of moderate size. Their use as a precision instrument is often neglected and the need for careful handling forgotten: familiarity breeds contempt. Too often one walks into a site office to find an instrument, still affixed to the tripod, propped up in a corner of the room! Such neglect not only increases the chance of damage, but also allows dust, condensation and other harmful elements to infiltrate the components of the levels. All levels are supplied with fitted cases, so why not use them! These cases provide stable and secure storage and a safe means of transport from office to site. Metal cases are particularly durable. Most cases are provided with a sachet of silica-gel crystals to absorb moisture and this should be replenished at regular intervals.

The level not generally be moved whilst affixed to the tripod. With the tripod legs over one's shoulder the instrument is in an unstable position and is vulnerable to vibrations and accidental damage by hitting obstructions behind. When transporting the instrument over a short distance where re-boxing is impractical, fold the instrument and tripod in a vertical position and move carefully to the next set up point. For longer moves the level should be re-cased.

To ensure that levelling is accurate the engineer must check his instrument regularly, to prove collimation is correct. This should also be done for new or hired instruments and when they are returned from service. The procedure for checking will be explained later in this chapter.

If collimation is found to be inaccurate it is not advisable, in the author's opinion, for the engineer to attempt his own level adjustment although some literature does advocate and explain the procedures involved. Faulty instruments should be sent to the service laboratory for proper repair.

Finally, keep the equipment and levels clean. During bad weather, maintain equipment indoors. Failure to follow these simple rules will inevitably result in errors and consequent waste of time and money – they should, therefore, become routine.

Theodolites

Most of the points discussed with regard to levelling instruments also apply to the theodolite. However, because it is a more complex machine and of heavier construction, some extra problems can occur. There are more controls, etc., so regular maintenance and cleaning is even more important. The method of setting up involves greater use of the adjustable legs of the tripod than is the case with the level (which may have the advantage of a ball and cup instrument mounting). This leads eventually to worn leg adjusting screws which, combined with movement by shrinkage of the timber tripod legs, can result in slippage. The only effective answer to such a problem is to replace the tripod – temporary packings between the locking screw and legs should not be used.

In most theodolites light enters the scale reading areas by means of an adjustable mirror, so it is important to ensure that the mirror and its opposing entrance glass are kept clean and free from damage.

If the instrument is of the type allowing interchange between tribrachs, take care to avoid damage to the fixing legs of the theodolite and check that no free movement is apparent.

E.D.M. equipment

Electromagnetic distance measurement equipment is expensive and requires careful storage and handling. Correct transportation is important, whether using the total – station type, or adapt-on type of E.D.M.

Connection leads to battery packs are almost always exposed and have to be manually connected. Care must be taken when assembling E.D.M. to avoid damage to lead connections and battery packs. Most manufacturers avoid the possibility of incorrect connection assembly by using different plug fittings to wires between components. When assembling adapt-on E.D.M. equipment (adapt-on describes the theodolite mounted types) the engineer must satisfy himself that correct adaption has been made – the connection should not be forced into place. The manufacturer's checking procedures should always be followed after setting up the equipment.

E.D.M. procedures involve the use of ancillary equipment such as extra tripods, prism reflectors, targets, extra tribrachs and support poles, etc., and these should also be handled with care and correctly stored, transported and cleaned.

Summary of equipment care

All optical equipment requires protection from inclement weather. Many manufacturers provide plastic covers to place over instruments set up on site but temporarily out of use during heavy rain, etc., but a plastic bag will also give the required protection.

It is obviously inadvisable to leave any instrument set up but unattended in the field where it may be stolen or accidentally damaged by machinery or other elements. It is not unknown for an instrument on site to be knocked over, damaged and re-erected by the offender. There may be no visible evidence of damage and the unsuspecting engineer transports the instrument to his next set up point unaware that he is using a faulty machine. All operatives should be asked to report such accidents as and when they occur.

Use of the equipment

Taping

When setting out, accurate taping is as important as levelling and theodolite work. It is pointless to spend money on extremely accurate levels and theodolites and then to destroy their results by inaccurate taping.

We have seen in Chapter 1, how a spirit level should be used to ensure a horizontal dimension is obtained over short lengths, and we have explained in this chapter, the maintenance and precautions necessary when using tapes. Many users over-pull the tape resulting in movement of the peg or nail at the hook end especially when the peg is located in soft ground. Always ensure that the tape is not snagged by an obstruction between points and that the band is in a straight line. On high accuracy work the use of a constant tension handle, or even an invar band (such bands have a low expansion value), may be required, and slope angles may have to be taken when tape lengths make the spirit level method too inaccurate.

The tape end or hook varies from manufacturer to manufacturer and will require a different allowance when hooked over a nail – usually the dimensions start from the very end of the hook. These allowances should be calculated to give the distance from the centre of the nail, depending on the tape and nail size used. However, when the tape is held over, or against, a position such allowances need not be made.

When taping for accurate work it is often necessary to make corrections for slope whilst the slope distance is taped, and the vertical angle between points is measured, the length of the slope is then adjusted by means of trigonometrical formulae. Likewise, corrections for temperature, scale factors and tension may also need to be made. Explanation of these correction procedures is beyond the scope of this book, but such corrections should be used when required. Before scale factors which require adjustment of the taped length, are applied (see Ch. 3) the engineer should check with the architect or resident engineer to see if such corrections are requested.

Setting out right angles by tape

Two tapes may, if desired, be used to set out right angles if a fixed base line is available. This method is only recommended where no side exceeds 30 m. Using the theorem of pythagoras for a right angled triangle as in Fig. 2.1 $z^2 = x^2 + y^2$ it can be shown that the ratio between the three sides can be $5^2 = 3^2 + 4^2$ or the commonly termed $3 : 4 : 5$ ratio.

Example

In Fig. 2.1 line A/B (or x) is a baseline 8.4 m long. To calculate the lengths of A/C (or y) and B/C (or z) the formula can be transposed as follows:

$$\text{Where } x = 8.4 \text{ m}$$

$$z \text{ can be seen to } = \frac{5}{3} \times 8.4 \text{ m}$$

and y can be seen to $= \dfrac{4}{3} \times 8.4$ m

Thus transposed z must $= \dfrac{8.4 \times 5}{3} = \dfrac{42}{3} = 14.0$ m

and y must $= \dfrac{8.4 \times 4}{3} = \dfrac{33.6}{3} = 11.2$ m

Once the values of z and y have been calculated, two tapes can be swung in an arc from points A and B until the two values intersect at point C.

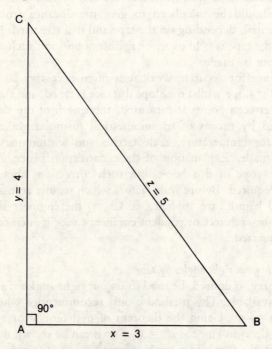

Fig. 2.1 3 : 4 : 5 ratio triangle

Precautions when taping
BS 5606 (1978) recommends the following taping procedures for improved accuracy.

 1 Full dimensions should never be built up from a series of intermediate dimensions, since this can generate cumulative errors. The tape should be extended over its full dimension and the intermediate dimensions marked from it.

2 Support the tape at intervals of 10 m.
3 Determine the tape temperature and apply correction if necessary.
4 Measure the slope length and apply correction if necessary.
5 Apply a compromise tension of 70 N when using a 10 mm wide steel tape and 105 N when using a 13 mm wide tape (this will enable the tapes to be used either fully supported or in catenary, with a resulting error due to sag and tension not exceeding ± 3 mm over 30 m). The appropriate tension may be applied using either a constant tension device or a spring balance.
6 Read the tape to the nearest millimetre, subdividing the 5 mm graduations by eye.

(*Note*: N is the abbreviation for Newtons, the metric unit of force.)

The site engineer must make his own judgement as to the number of precautions he should adopt. This will relate to the accuracy required and the distances being measured on the setting out in hand.

Levelling

Once the engineer has familiarized himself with the operation of his instrument, whether dumpy, quickset, precise or automatic type, the principles for use remain similar. The main variations between level types occur in the method of adjustment to render the instrument in a level setting.

Setting up

The first priority when setting up the instrument on site is to ensure that the tripod is positioned on a secure base with the legs firmly planted in the ground. If the ground is particularly soft and muddy, movement of the tripod legs may occur as the engineer walks around the instrument, this in turn results in the instrument moving out of level. When such conditions are unavoidable, the operator must be cautious in his approach to and movements around the tripod.

If the level is to be set up on a concrete or similar smooth hard surfaces, care must be taken to ensure the tripod does not slip after the instrument has been levelled. Where such set up points are frequently required, it may be useful to purchase a tripod base, made by many manufacturers, which allows the tripod to be set up safely on a slippery floor.

Avoid setting up a level on tarmac during hot weather, as the weight of the instrument may force the legs to penetrate into the material, thus lowering the collimation of the instrument.

Make certain in all cases that the instrument is correctly fixed to the tripod.

Levelling the telescope

Dumpy level The telescope of a dumpy level has a bubble tube set alongside. The instrument is levelled by means of a three footscrew base, the tube bubble is turned parallel to two screws and these screws are then turned away from, or towards, each other until the bubble shows level. The bubble is then turned at right angles to these two screws so that the telescope points over the third screw, and adjustment is made by this single screw until the bubble again shows centre. This sequence of operations is then repeated over the other screws until the bubble shows constantly level when the instrument is slowly turned through 360°. Readings are then taken and no further adjustment of the instrument is required until the instrument is moved and re-sited. Figure 2.7 shows the practice of plate bubble levelling in plan, and lists the operations.

Tilting level The tilting level is an improvement on the design of the dumpy level as it provides movement of the telescope by a small amount in the vertical plane. Movement is achieved by resting the telescope on a pivot. With a tilting screw, generally located below the eyepiece end of the instrument, the telescope can be adjusted for level.

The instrument is first roughly levelled using a centre circle bubble mounted on the tribrach using three footscrews. Final levelling is then carried out before each staff reading is taken. The tube bubble fixed to the telescope is adjusted by moving the tilt screw until the bubble settles in a central position.

Quickset level This type of instrument has a ball and cup mounting rather than a three footscrew base which gives the user the facility for fast rough levelling by means of a circular bubble. Once roughly levelled, however, its method of operation is the same as that of a tilting level, i.e. after approximate levelling the

30

instrument is finely adjusted for each reading by means of a tilting screw, which adjusts a tube bubble fixed to the telescope.

Precise or engineers level The precise level or, as some manufacturers now call it, engineers level is set up in the same manner as the tilting level, but gives a higher degree of accuracy because it uses a coincidence bubble rather than a tube bubble. The bubble is adjusted by means of a tilting screw which is usually on the side of the instrument. (See Fig. 2.2 for details of coincidence bubble.) Adjustment of the coincidence bubble is made each time a reading is taken.

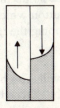
Position 1.
Bubble sides do not coincide and require adjustment in direction of arrows

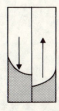
Position 2.
Adjustment made too great and bubble still shows that instrument is out of level

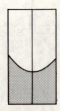
Position 3
Bubble sides coincide showing that instrument is level

Fig. 2.2 Example of coincidence bubble operation

Automatic level The automatic level now available gives the same high degree of accuracy as the precise level but is faster to operate. The risk of forgetting to re-level the bubble each time the reading is taken is eliminated. After levelling a circular bubble using the three footscrews, the instrument is self levelling by means of a pendulum prism suspended within the casing and this sets the line of sight horizontal for all pointings of the telescope. The incorporation of pneumatic damping insulates the pendulum from the influence of strong winds, traffic vibrations, etc. and ensures a stable line of sight.

Occasionally prisms have been known to stick, making the line of sight out of true. As a precaution, the engineer should look through the telescope and **lightly** tap the leg of the tripod. If the prism is not stuck the view through the lens via the prism will move and levelling procedures may commence. Some manufacturers now incorporate a compensator control button which when pressed moves the prism, i.e. the same test as tapping the tripod.

Reading the level

When the instrument has been levelled the user will see a horizontal line of collimation through the telescope tube. Etched onto a reticule in the lens of the instrument are vertical and horizontal lines. The usual pattern for such lines is shown in Fig. 2.3. The

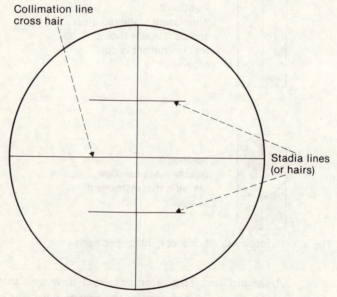

Collimation line
cross hair

Stadia lines
(or hairs)

Fig. 2.3 Example of etched cross hairs on reticule located in the diaphragm of a level telescope

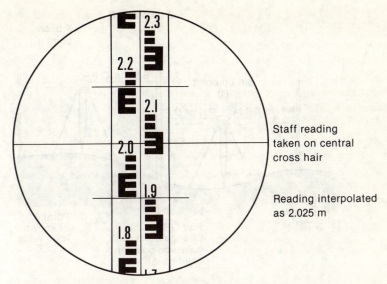

Staff reading
taken on central
cross hair

Reading interpolated
as 2.025 m

Fig. 2.4 Example of staff reading via instrument

line running horizontally across the centre of the reticule is for taking the staff reading, giving the level value. The two smaller horizontal cross lines above and below the centre are known as the stadia hairs and are used for distance measurement. (Fig. 2.4 shows the staff reading when seen through the lens.)

After setting up, the engineer must find out the level height of his instrument by reference to a datum, usually from a T.B.M. (temporary bench mark) or ordnance survey level point. The height of the instrument will then be known as the collimation height. The datum, if tied into the ordnance survey will be at a known value above mean sea-level (mean sea-level having been set by the O.S from a base at Newlyn in Cornwall). A datum point may, however, be referenced to a particular point on the site and given an arbitrary value from which all levels are derived, although it is better practice to refer to the O.S. values whenever possible.

Once the height of collimation has been established, further staff readings are taken – these readings being deducted from the collimation height to give the value of the point levelled. Figure 2.5 shows the principles of level reading in diagrammatic form. As a precaution it is advisable to keep all staff readings within a 60 m radius of each set up point.

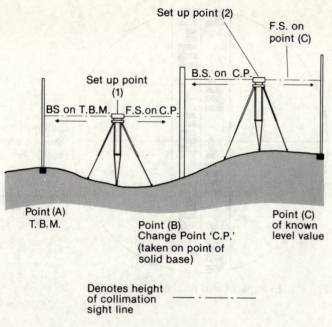

Set up point (2)

F.S. on
point (C)

B.S. on C.P.

Set up point
(1)

BS on T.B.M.   F.S. on C.P.

Point (A)
T. B.M.

Point (B)
Change Point 'C.P.'
(taken on point of
solid base)

Point (C)
of known
level value

Denotes height
of collimation —— · —— · ——
sight line

Fig. 2.5 Diagrammatic example of level reading

Method

1 Instrument set up at point (1)
2 Staff B.S. taken on T.B.M. point (A)
3 Height of collimation at point (A) = B.S. staff reading + R.L. of T.B.M.
4 Staff F.S. taken on C.P. (B)
5 F.S. reading subtracted from height of collimation to find R.L. of C.P.
6 Instrument moved to set up point (2)
7 B.S. taken from point (2) on C.P. (B) used as datum
8 Height of collimation at point (2) = B.S. reading + R.L. of C.P. (B)
9 Staff F.S. reading taken on point of known level at (C)
10 F.S. reading at (C) subtracted from H. of C. at (2) to give R.L. of point (C)
11 Checks on arithmetic made and comparison of R.L. at point (C) made with known level

Level booking

It is preferable that all calculations, when levelling, are carried out in a levelling field book. There are two usual methods of booking: the collimation method and the rise and fall method. The author prefers the collimation method as this involves less complicated booking. Checks should be made at the end of each section of levelling to a known datum point, thus making apparent any errors which may have resulted during the period of levelling. Example of booking for both methods are shown in Fig. 3.1 in Chapter 3.

Change points (C.P.)

At some time during the levelling survey the engineer may be forced to move the instrument, as the fully extended staff may have dropped below the collimation height. This often happens on sloping sites. Before moving the instrument the engineer must ascertain whether a datum point of known value will be visible, if not a change point (C.P.) will be required from which the survey may continue, with reference to the datum already use.

A C.P. is merely a means of establishing a temporary level on a solid point for use as a datum when the level is re-sited. The engineer selects a firm location and whilst his assistant holds the staff on this spot the engineer takes the reading and moves the instrument to a new location. By deducting the reading from the collimation height, he can calculate the value of the spot at which the staff is held. A further reading is then taken on the staff from the instrument's new location, this in turn is added to the calculated datum level of the C.P., to give the new collimation height of the level. An example of such a calculation is shown in Fig. 3.1 in Chapter 3.

In order to ensure a high degree of accuracy on such staff readings, the staff man should face in the direction of the level and move the staff slowly to and fro. The lowest reading of the arc shows the engineer the point at which the staff is held exactly vertical.

Accuracy of readings

From the check made in his survey book the engineer will know the degree of accuracy obtained during his traverse. BS 5606: 1978 published by the British Standards Institution, London, advises that the following accuracy should be sought.

Site surveys For all types of survey the accuracy of level values should be:

    1 site T.B.M., relative to ordnance bench mark (B.M.) ± 10 mm
    2 spot levels relative to site T.B.M., ± 10 mm.

Note: On hard surfaces 90% of (2) should be accurate to ± 5 mm.

Setting out Table 4 of BS 5606 states the deviations that may occur and which will rarely be exceeded provided good practice is followed and reasonable care taken. Where possible, sight lengths should be made equal. For levelling, this table specifies the following deviations:

1 using builders class level	± 5 mm per single sight of up to 60 m
2 using engineers class level	± 3 mm per single sight of up to 60 m ± 10 mm per km
3 using precise class level	± 2 mm per single sight of up to 60 m ± 8 mm per km.

The engineer must instruct his assistant to hold the staff vertical at all times. Many staffs incorporate a circular bubble in their lower section to assist in achieving this, however, as shown under change point readings, an alternative is to move the staff slowly backwards and forwards and take the minimum reading. Always ensure that joints in the staff are properly connected.

Locating temporary bench mark
At the begining of a contract it will be necessary to locate within the site boundary a datum of known value. Where this datum is to be related to the ordnance survey grid, levels should be transferred from the nearest ordnance bench mark (B.M) to a temporary bench mark (T.B.M.) on site. Ordnance bench marks are denoted by the sign shown in Fig. 2.6(a) and are located on solid sites such as churches, houses and bridges, etc. The level value of the B.M. is recorded on ordnance survey plans, which are available at their local offices, stationers, or are specified by the architect. The reduced level value is referenced to the horizontal line on the B.M. sign, usually etched into the material of the structure. The T.B.M. must be located on a stable base and in a permanent location which will not be disturbed during the course of the works. A square slab

of concrete dug into the ground to a safe depth will serve this purpose. The slab should have either a metal plate or stout nail cast into and slightly protruding above its surface, upon which the level reading can be taken. On large sites it will be necessary to locate more than one T.B.M. within the site area – these often being placed on grid stations, (see Ch. 4).

When transferring the level from the B.M. to site, endeavour to keep foresights and backsights equidistant, to reduce the possibility of error.

Checking the accuracy of the instruments

It has already been stated that periodic checks on the accuracy of the level must be made on the site. Such a check must always be made if there is evidence, or even a suspicion, that the instrument has been knocked or damaged. The procedure for checking is as follows (with reference to Fig. 2.6)

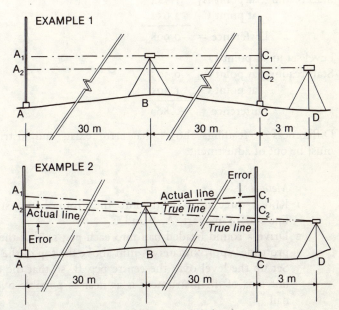

Fig. 2.6 Checking the accuracy of the surveyor's level

Fig. 2.6(a) Ordnance Bench Mark.

Example 1
Level set up at point B
Staff reading: at point A_1 1.735
          at point C_1 2.643

Difference = −0.908

Level set up at point D
Staff reading: at point A_2 0.486
          at point C_2 1.394

Difference = −0.908

Differences in readings for both set up points agrees – therefore instrument is in adjustment and no error exists

Example 2
Level set up at point B
Staff reading: at point A_1 1.744
          at point C_1 2.652

Difference = −0.908

Level set up at point D
Staff reading: at point A_2 0.505
          at point C_2 1.395

Difference = −0.890

Differences in readings shows that error exists and instrument must be out of adjustment.

Method

1 Knock three pegs firmly into the ground, in a straight line and 30 m apart, A, B and C.
2 Drive a round headed nail into each peg so that the head protrudes approximately 5 mm above the top of the peg.
3 Set up the level over the centre peg B, so that the centre line axis of the instrument is as near as possible over the nail.
4 Instruct an assistant to hold the staff carefully on top of the nails on pegs A and C respectively. (To gain an accurate vertical reading the staff should be moved slowly to and fro.)
5 Take staff readings on points A and C and calculate the difference.

6 With the staff still held on point C, move the instrument 3 m past C still in line with A, B and C.

7 Again take staff readings on points C and A from the level's new location and calculate the difference.

If the instrument is accurate (i.e. the line of collimation is in a horizontal plane) then the difference between the readings will be identical from both set up points. However, where a faulty level is used the difference will vary because the line of collimation will be untrue and inclined to the horizontal. In Fig. 2.6 the errors are exaggerated to show the variations more clearly.

Adjusting for parallax

Staff reading errors may result if the instrument is out of parallax. The engineer must test the focus of the level for parallax due to variations of eyesight between users.

It is useful to start by adjusting the eyepiece until the crosshairs are distinct and in sharp focus. This may be achieved whilst viewing a light background such as the sky or a piece of white card. Looking through the telescope move the eye slightly above and below the crosshair and if the line seems to fluctuate adjust the focus until the movement stops. If this alteration of focus blurs the staff, adjustment of the eyepiece to clarify the sighting should be made and the crosshairs will appear more vivid. The procedure should be repeated until the crosshairs remain stable on the object when the eye is moved, and the focusing is clear.

Summary of terms frequently used during levelling

Collimation	This is the term that denotes the imaginary horizontal line visible through the object lens via the central cross hair of the level.
Backsight (B.S.)	The first reading taken from the instrument, usually on the datum.
Foresight (F.S.)	The last sight read before the instrument is re-positioned.
Intermediate sight (I.S.)	The readings taken other than backsight and foresight.
Temporary bench mark (T.B.M.)	A temporary datum fixed on site for use as a basis for level values, reference and checking point.

Ordnance bench mark (B.M.)	Levels of known value establish-ed by the ordnance surveyors on various solid locations through-out the country. Related to mean sea level value at Newlyn in Cornwall.
Reduced Level (R.L.)	The finished height calculated from the selected datum.
Datum	A point of initial reference of a given level value having a solid base.

Theodolites

A theodolite will be used both to set out and to take vertical and horizontal angles, where the accuracy of 3 : 4 : 5 taping and dumpy level scales are not adequate. The theodolite is now commonly used for all angle measurement unless space and the project dictate otherwise.

Setting up

Setting up of the theodolite is similar to that of the three foot-screw dumpy level, but is more complex because the instrument must be set with its vertical axis over a known point. Before the advent of the optical plummet the instrument was located over its set up point by means of a plumb-bob. Some older instruments and also some users of modern equipment still use this method, although it is less precise and more difficult to carry out especially in windy conditions.

Plumb-bob method To set up using a plumb-bob the en-gineer must roughly locate the tripod so that its centre is approxi-mately above the set up nail or mark. After attaching the plumb-bob to a hook hung below the tripod plate screw, the tripod is moved and adjusted on each leg until the plumb-bob is as close as possible to the required set up location. With the head of the tri-pod approximately level, the theodolite is then attached and accu-rately levelled. After levelling of the instrument the plumb-bob is again inspected, and if it is off centre, an adjustment is made to the tripod, or if only a small adjustment is required it is made by slackening the tripod fixing screw and sliding the instrument over on the tripod head. This procedure may have to be repeated several times before the instrument is truly level and the plumb-bob exactly vertical over the required spot.

On some types of instrument there is a shifting plate between the alidade and the tribrach which allows the instrument to be moved to adjust for slight discrepancies without (or only slightly) affecting the level of the theodolite.

Optical plummet method Most modern theodolites have an optical plummet frequently set in the alidade of the instrument. With this component an alternative method of setting up can be used which allows faster and more accurate location of the theodolite. This second method of set up does away with the necessity for a plumb-bob (unless desired in tripod centering). Equipped with such an instrument, the engineer can erect his theodolite in the following manner.

Firstly stand the instrument approximately over the point of location. Grasp and raise slightly two of the tripod legs and oscillate the instrument whilst viewing the mark through the optical plummet. Once approximately over the mark, lower the legs and push each one into the ground with equal force. Check the centering of the mark through the optical plummet and adjust until centred by means of the three footscrew tribrach. Next take each tripod leg in turn and revolve the theodolite until its plate level tube is parallel with the line of the leg. Unscrew the leg clamp and slide the leg up and down until the tube bubble is central, then tighten the clamp. Repeat the process for each leg. The theodolite must now be levelled by means of the three footscrews as explained in Fig. 2.7. Check through the optical plummet to ensure that the instrument is over the mark. If only slight adjustment is required slide the instrument over on the tripod head or use the shifting tribrach, if provided. When adjustments are made, as is normally the case, the instrument must of course be re-levelled and the procedure repeated until the theodolite is both level and over the set up mark.

Controls on the theodolite

Controls obviously vary between different models but the principles for use are similar.

The controls for a theodolite are shown in Fig. 2.8: the example used is the Sokkisha TM20E/ES. The 20E has a detachable

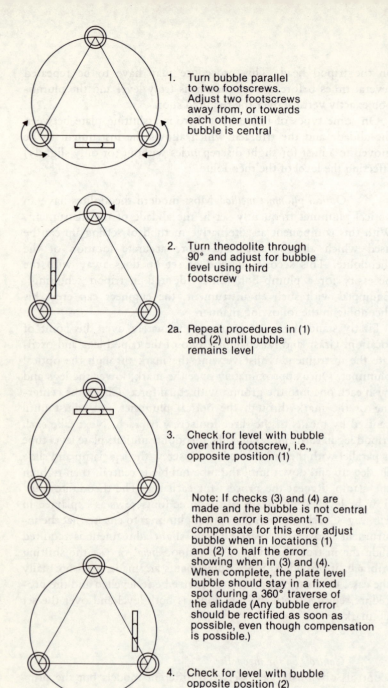

1. Turn bubble parallel to two footscrews. Adjust two footscrews away from, or towards each other until bubble is central

2. Turn theodolite through 90° and adjust for bubble level using third footscrew

2a. Repeat procedures in (1) and (2) until bubble remains level

3. Check for level with bubble over third footscrew, i.e., opposite position (1)

   Note: If checks (3) and (4) are made and the bubble is not central then an error is present. To compensate for this error adjust bubble when in locations (1) and (2) to half the error showing when in (3) and (4). When complete, the plate level bubble should stay in a fixed spot during a 360° traverse of the alidade (Any bubble error should be rectified as soon as possible, even though compensation is possible.)

4. Check for level with bubble opposite position (2)

Fig. 2.7 Levelling procedure for level or theodolite plate level on three screw tribrach

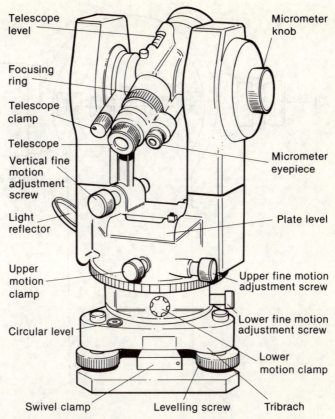

Fig. 2.8 Controls layout on typical theodolite (Sokkisha TM20E/ES)

tribrach and the 20ES a shifting tribrach for centering over a mark through an optical plummet.

The cross sections through the Wild T2 theodolite shown in Figs. 2.9 and 2.10 detail the optical light paths through the instrument and show something of the complex construction.

Before setting up a theodolite on the tripod, adjust each of the footscrews to the centre of their run, this will allow maximum transit in each direction for levelling purposes.

Reading the angles

Once the theodolite is set up, the engineer must obviously be fully conversant with the controls of the equipment in order to use it. And as with any instrument, it is essential to read the manual.

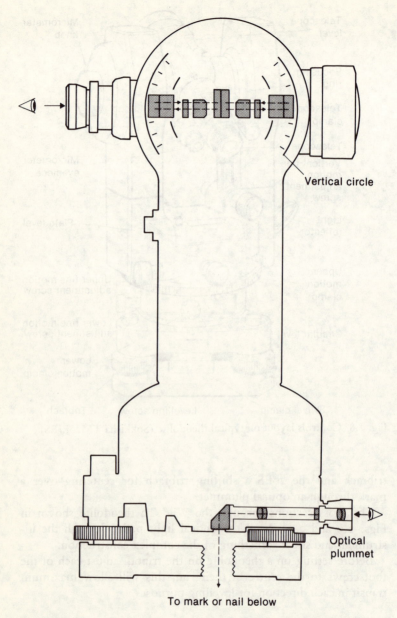

Vertical circle

Optical plummet

To mark or nail below

Figs. 2.9 and 2.10 (opposite)   Optical light paths through Wild T2
theodolite

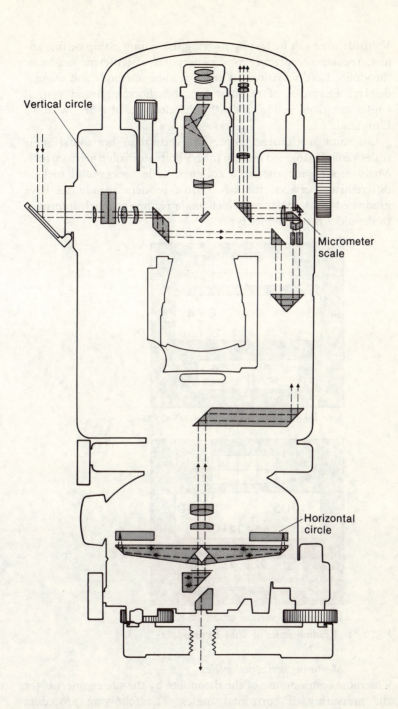

Vertical circle

Micrometer
scale

Horizontal
circle

45

Valuable time can be lost by moving the wrong clamp or fine adjust, regular site practice is very important. Different makes of theodolite have variations in angle reading methods and angular display. Examples of readings for both horizontal and vertical angles are given in Fig. 2.11: the theodolite used is the Wild T2 Universal.

On some instruments either the vertical or horizontal angle scales can be blanketed out, on others both are visible permanently. Many models incorporate a coloured scale background to help differentiate between the two. Most modern theodolites have graduated glass circles on which angle reading is much improved to the older metal vernier types.

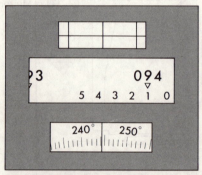

360° reading: horizontal circle 94° 12′ 44″

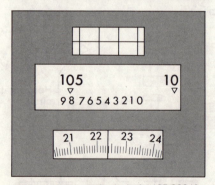

400° reading: vertical circle 105.8224°

Fig. 2.11   Reading scales of Wild T2 theodolite

Measuring horizontal angles
The most common use of the theodolite by the site engineer is for the measuring of horizontal angles. The following procedure

46

should be adopted and Fig. 2.8 referred to for operating control. The angle will either be measured from a point of known bearing or, alternatively, from a point of assumed bearing usually taken, for convenience, as 00° 00′ 00″ (or 360° on some instruments). The instrument is set up, the horizontal scale is viewed and the instrument is either locked in position on a preset bearing value, or fixed at zero. To set the scale the lower motion clamp is locked and the theodolite turned until the desired angle is met, then the upper motion clamp is affixed. Fine adjustment is made using the micrometer knob, if one is fitted,[1] and the upper fine motion adjustment screw. Check that the focus of the instrument is clear and adjust for parallax. With the upper motion clamp still fixed, release the lower motion clamp and sight the target, lock the lower clamp and make any fine adjustments on the lower fine motion adjustment screw.

To set out or sight the next object, release the upper motion clamp and swing the theodolite until the desired point or angle is approximately met. Focus the telescope and accurately adjust to the required location by means of the upper fine motion adjustment screw. Locate the point or record the angle and repeat for further locations as desired, the final reading being taken on the point of origin as a partial check for the accuracy of the traverse. The points are usually checked, especially on work requiring fairly high accuracy, by changing the face of the instrument (i.e. transiting the telescope through 180°), and re-sighting the points previously taken. If the instrument is in complete adjustment the difference between the new and previous readings will of course be 180°. However, slight errors often prevail once angles have been read on face left and face right. When the point of origin is re-sighted and the angle does not coincide with a 180° difference, the instrument should not be adjusted, but the reading should be logged as viewed. Checks of the other locations are then made, the instrument being revolved in the opposite direction to that on the first traverse.

For readings taken on fixed existing locations, both readings (i.e. face left and face right) are logged, the mean angle then found and recorded (see p. 63). For locating points on the site, marks are made from both faces then bisected to position the final location. For surveying purposes it is often necessary to obtain further readings of these points, by another one or more round of angles, the

---

[1] Depending on scale reading type, a micrometer screw may or may not be fitted.

theodolite scale being adjusted to say 90° when sighting the point of origin to give an even more thorough check.

Measuring of vertical angles

The engineer will have more cause to measure horizontal than vertical angles, unless he becomes involved in trigonometrical levelling or slope angle measurement. Vertical angles are measured as follows:

Establish the height of the instrument by measurement to its trunnion axis and log this together with the height of the target used.

Set the altitude bubble (if fitted) to the centre of its run before each reading is taken. Commence readings with the instrument face left and repeat the round of angles with the readings taken with the instrument face right.

Many modern theodolites incorporate an automatic index which makes levelling of an altitude tube bubble obsolete. The vertical angle measurements can be thus taken faster and just as accurately once the instrument is levelled by the plate level. With the instrument level and finely adjusted on to the angle, or spot, to be located, either mark the object or read the angle by reference to the horizontal central cross hair. Change the face of the instrument and repeat the procedure to establish a mean reading.

As explained in Chapter 3, when booking it may be necessary to note if the angle is in elevation or depression, depending on the divisions of the vertical circle and to record the exact point read or set out.

Any altitude/telescope bubble provided must be levelled before each reading is taken. The time saving achieved when using an instrument with an automatic index can therefore be appreciated.

Checking the theodolite

Plate level The plate level is partially checked each time the theodolite is set up as, after levelling, the bubble should remain central wherever the instrument is pointed. Any errors in the plate level should be rectified or reading accuracy will be impaired, although adjustments can be made to compensate for errors when setting up the instrument (see dumpy level and Fig. 2.7).

Horizontal collimation Set up the instrument over a nail driven into a solid peg. Sight in another peg (A) 50 m away and accurately locate another nail on line, the instrument should have the lower and upper plates clamped and the horizontal scale on zero. Transit the telescope and sight another peg (B) 50 m away in the opposite direction and mark on line and distance in pencil. Unlock the upper motion clamp and resight (A) with the instrument on the opposite face to that previously used. Fix the upper motion clamp, transit the telescope and re-sight (B). If the second sighting corresponds with the mark previously made, the instrument is accurate. If the marks do not agree then the theodolite requires adjustment to rectify the error. The difference between the two marks at (B) represents four times the collimation error, with the mid point being the true line.

Transit (trunnion) axis The transit axis should be perpendicular to the vertical axis of rotation and thus horizontal when the instrument is set up and adjusted to a level state. On instruments out of adjustment, the theodolite telescope will not transit in a vertical plane, making angular measurement incorrect. To test the transit axis the following procedure should be followed.

Set up the theodolite at a spot from which a high point is visible upon which an accurate sighting can be made e.g. the roof ridge fascia board apex on a gable some 25 m in height. The distance from the building for the set up should also be approximately 25 m to give a 45° angle of incline. After sighting an exact location on the high point depress the telescope and mark the wall at the ground level of the building. Alter the face of the instrument and repeat these operations, thus giving a second mark on the wall. If both marks coincide, the instrument is accurate, if they are apart there is an error. If this error exceeds 4 mm over the 25 m distance the instrument should be sent away for adjustment.

Optical plummet check (alidade mounted plummets) Secure a piece of card below the central vertical axis of the instrument and sight a mark onto this through the optical plummet. Rotate the instrument alidade through 180° and again mark the card, thus the plummet is viewed from the opposite side than when making the first mark. If both marks coincide, the plummet is accurate, if they are apart the bisecting of the lines will show the true central plummet axis. However, adjustment must be made to avoid errors in work on site.

Vertical circle index Procedures for checking this are not included in this book although methods are available. The vertical circle is used more for surveying purposes than in setting out, and as such, is little used by the engineer. Only where slope angles are taken on a steep site requiring long tape lengths, or on occasional survey work, will the vertical circle be used. The techniques for checking can be found in surveying textbooks.

Altitude bubble check Testing procedures for the altitude bubble are not explained in this book as many modern theodolites have an automatic index and are thus plate levelled for vertical circle reading. However, when testing is desired, on instruments incorporating an altitude bubble the procedure can be found in many reputable surveying books.

Precautions to be taken when using a theodolite

1 Do not overtighten clamping screws on the instrument.
2 Avoid knocking other controls when sighting and focusing through the telescope.
3 Never carry the instrument by the telescope, but always on a solid point such as the telescope standard.
4 Always adjust for parallax before sightings commence.
5 Ensure the theodolite is correctly attached to the tripod and accurately levelled when set up.
6 Use a theodolite of equal or greater accuracy than the work to be undertaken requires.
7 On important and especially long sights take readings on two faces of the instrument, taking the mean reading as correct.
8 Never, unless totally unavoidable, sight a long line from a short baseline – always the opposite.
9 Avoid flat, i.e. less than 30°, intersecting angles when setting out major control stations.
10 Ensure a stable set up point is established.

Suggested permissible deviations using the theodolite

BS 5606: 1978 and BRE Digest 202: 1977 denote that the accuracy in use of measuring instruments should not exceed those shown in Table 2.1.

Electromagnetic distance measurement (E.D.M.) equipment

Electromagnetic distance measurement (E.D.M.) can be used by the engineer to establish all the positions of critical accuracy such

Table 2.1 Suggested permissible deviations using the theodolite from BS 5606:1978 and BRE digest 202:1977

Measurement	Instrument	Deviation	Comment
Angular (A)	Glass arc theodolite (with optical plummet or centering rod) Reading directly to 20″	± 20″ (± 5 mm in 50 m)	Scale readings estimated to nearest 5 seconds. Mean of two sights with readings in opposite quadrants of the horizontal circle, one sight on each face.
Angular (B)	Glass arc theodolite (with optical plummet or centering rod) Reading directly to 1″	± 5″ (± 2 mm in 80 m)	Mean of two sights, one on each face with readings in opposite quadrants of the horizontal circle.
Verticality	Glass arc theodolite as (A) above but fitted with diagonal eyepiece	(± 5 mm in 50 m)	Mean of at least four projected points each one established at a 90° interval around a horizontal circle. Good illumination required inside buildings
Verticality	Glass arc theodolite as (B) above but fitted with diagonal eyepiece	(± 5 mm in 100 m)	As above.

as grid points, road intersection points, building corner locations and drainage positions on a site. Compared to more established methods E.D.M. has the benefit of improved accuracy especially over long distances. The range of makes, types and manufacturers now available satisfies the requirements of most users. Chapter 1 gives some guidelines for selecting E.D.M. equipment.

An ever increasing range of instruments is available and it is not practical to instruct on their operation as all have different controls and facilities. The main types used are explained briefly in Chapter 1, and page 25 indicates the precautions to be considered when handling E.D.M. Most types on the market at present offer the facility of slope distance measurement. However, more comprehensive machines can give readings for slope distances, horizontal distance and the polar co-ordinate differences merely by feeding in (usually by push button) the angles read from the theodolite. Total station types have electronic digital display of angles and can automatically calculate slope, horizontal and co-ordinate lengths from mounted manual controls. It is possible to record the readings taken in the field directly onto paper print outs or recording tape. Electromagnetie distance measurement offers particular advantages on steeply sloping sites where conventional measuring methods would be difficult and time consuming. Where rivers have to be crossed and in many applications where normal taping or optical distance measurement would prove difficult E.D.M. can be useful.

Programmable calculator operation

A general explanation of the operation of programmable calculators is given because each make of instrument is different. The facility to program should be used whenever practical, and is extremely useful for road and co-ordinate calculations. The operator must familiarize himself with the calculator's operation in a programming mode but any programs compiled will need careful construction and testing. Before formulae can be entered into a program the operator must consider if transposition is required, and should adopt the same approach as the calculator will adopt to resolve the problem.

Example

Take a simple formula such as that for a deflection angle calculation often used in roadworks:

$$\theta = \frac{L}{R}$$

Where θ = deflection angle in radians
L = arc length
R = radius

As the program may need to be 'run' ('run' is the term used to describe a program being set to work) several times, it must be compiled so as to allow varying values to be given for both L and R. It is usual, therefore, to use the calculator's memories to store different values and the instruction in the program will be to recall these memories as they are required.

The following program has been devised for use with the

Texas TI57 calculator which is referred to repeatedly throughout this book. For other brands and types of calculator the programming examples that follow will obviously need to be changed.

Formulating a program for Texas TI57 (Fig. 1.3.)

Firstly transpose the formula to enable various values to be given to L and R and as this involves memory recall, decide into which two memories L and R will be inserted. Say memory 1 for L and memory 2 for R, denoted on the calculator as STO (store) 1 and STO 2. Any values for L and R can be entered into STO 1 and STO 2.

Obviously before θ can be found, $L \div R$ must be evaluated, therefore the first steps of the program will be to RCL (recall) STO 1 and divide this by RCL 2. The result after applying the equals symbol will be θ given in radians.

The next stage of the program will be to convert θ to degrees, minutes and seconds, (D.M.S.), which can be read direct from the TI57 display. (*Note*: If further steps in the program require the use of θ in radians, the value should be stored in another memory.) Conversion to D.M.S. from radians can be made in several ways, but for the purpose of this example, θ has been multiplied by 180 and divided by π (3.141 592 7), then converted from decimal degrees to D.M.S.

On the TI57, before entering the program, the LRN (learn) button is pressed, putting the unit into learn mode. The program is completed by entering the R/S (run and stop) button, upon reaching this instruction the calculator will display the calculated result. The calculator is then taken out of learn mode by again pressing the LRN button and then pressing RST (reset) to commence the program at 00 (step 1).

A compiled program is shown in Fig. 2.12 and this would be entered into a calculator as indicated from each step number. When entry is complete, values for L and R can be placed in the memories and the R/S key pressed to run the program and resolve the problem.

Of course this simple example only outlines the calculator's capabilities and in later chapters more comprehensive programs are illustrated. The TI57 is capable of taking up to 50 program steps, and with use of its eight memories, facilities for sub routines and other scientific keys it is a valuable aid for the site engineer. Even more advanced calculators are available, these have larger program

Program record sheet for Texas T157 calculator

Program title:

To calculate the deflection angle of a curves arc.

Program drawn up by: R.W. Murphy   **Date:** July 1981

Formulae used: $\theta = \frac{L}{R}$ (note θ result in radians)

Program description and notes

Program shows θ in final result in degrees, minutes and seconds (D.M.S.) rather than radians, generally preferable when angle is required on site for establishment by theodolite. The angle can be read directly from the calculator display, the digits to the left of the decimal point being the degrees the two pairs after the decimal point, i.e., the first four figures to the right, being the minutes and seconds respectively.

Step number	Press	Key code	Calculator display	Step number	Press	Key code	Calculator display
1	LRN	–	∞ – 00				
2	RCL 1	33 – 1	01 – 00				
3	÷	45	02 – 00				
4	RCL 2	33 – 2	03 – 00				
5	=	85	04 – 00				
6	×	55	05 – 00				
7	1	01	06 – 0.0				
8	8	08	07 – 00				
9	0	00	08 – 00				
10	÷	45	09 – 00				
11	2nd π	30	10 – 00				
12	=	85	11 – 00				
13	Inv 2nd D.ms	– 26	12 – 00				
14	R/S	81	13 – 00				
15	LRN	–	0				
16	RST	71	0				

Information required in memories fed into:
STO 0 – ✓
STO 1 – Enter value of L in metres
STO 2 – Enter value of R in metres
STO 3 – ✓
STO 4 – ✓
STO 5 –
STO 6 –
STO 7 –

Fig. 2.12   Program record sheet for calculation of deflection angles on curves

and memory banks and as always the instruction booklet should be carefully read to ascertain all the unit's capabilities. Pocket computers are also now available for on-site use.

Laser levelling equipment

The use of laser levelling equipment is described in Chapter 1. Once set on a base of known datum, most laser levelling beam instruments are self adjusting to give a horizontal beam of light. The beam of light can then be revolved in a circle at varying speeds to a known height of collimation from which staff readings can be taken using either conventional or detector staffs The range of the instruments and their accuracy varies with type and site conditions. Drainlaying lasers were also discussed in Chapter 1.

3

Booking and recording of information

Far too often untidy booking or calculations carried out on scraps of paper cause mistakes of computation which lead to errors on site. Errors and subsequent time consuming research are often caused by failure to undertake checks both during and at the end of calculations and by not booking level readings. Many site offices are littered with scribbled calculation sheets, haphazardly stored in drawers and on work surfaces. The loss of time and patience caused by such filing methods is considerable and should not be tolerated. Logical filing and booking systems must be used.

This chapter examines those areas of work which should be recorded and shows examples of record sheets and charts. The reader may want to adapt the methods to suit his particular needs. Wherever possible new sheets should be devised to record other valuable information.

Methods of level booking

Height of collimation level book

The ruled columns used in a height of collimation type level book will be similar to those in Fig. 3.1. The first column backsight (B.S.), the second column intermediate sight (I.S.), and the third column foresight (F.S.), together with the remarks column, are the sections used to enter readings and observations taken in the field. The height of collimation and reduced level (R.L.) columns will be computed from the readings when the survey has been completed, either in the office or on site. This procedure applies when field surveying from a known datum point. However, when setting sight rails and drain inverts etc., from levels given on the drawing, the following method will apply.

After reading the backsight (B.S.) on the T.B.M., the height of

Collimation method

B.S.	I.S.	F.S.	Height of collimation	Reduce level	Remarks
1.455			100.015	98.560	Station A T.B.M.
	1.335		(×4)	98.680	Top of peg 1
	1.680		(400.060)	98.335	Top of peg 2
	0.995			99.020	Top of peg 3
1.795		0.660	101.150	99.355	Change point peg 4
	1.330		(×2)	99.820	Top of peg 5
		0.665	(202.300)	100.435	100.483 Station B T.B.M
3.250	5.340	1.325	602.360	595.695	
1.325	5.340				
1.925		595.695		1.925	(reduce level at) (station B shows) (field error of)
		602.360			(0.002mm to known) (T.B.M. value)

Rise and Fall method

B.S.	I.S.	F.S.	Rise	Fall	Reduce level	Remarks
1.455					98.560	Station A T.B.M
	1.335		0.120		98.680	Top of peg 1
	1.680			0.345	98.335	Top of peg 2
	0.995		0.685		99.020	Top of peg 3
1.795		0.660	0.335		99.355	Change point peg 4
	1.330		0.465		99.820	Top of peg 5
		0.665	0.665		100.435	100.483 Station B T.B.M
3.250		1.325	2.270	0.345	98.560	
1.325			0.345			
1.925			1.925		1.925	(reduce level at) (station B shows) (field error of) (0.002mm to known) (T.B.M. value)

Fig. 3.1 Examples of 'collimation' and 'rise and fall' methods of level booking

collimation will be calculated and entered in to the appropriate column. From the drawing, the values of the object to be levelled will be entered into the reduced level column, these being deducted from the height of collimation to give the required intermediate sight (I.S.) values. These values give the staff readings required to position the object's height.

It can be seen from the worked example in Fig. 3.1 that the B.S. is the first reading taken on the T.B.M., following readings then being entered in the I.S. column, until the final reading, which is recorded in the F.S. column, is taken prior to moving the instrument.

When transferring the instrument by use of a change point (C.P.), the preferred method of booking is also shown in Fig. 3.1. It can be seen that the reading on the C.P. (being the last reading before the level is re-sited) is entered in the F.S. column, and the reduce level (R.L.) of this point calculated to form a C.P. datum. Once the instrument is re-set and ready for use, a B.S. taken on the C.P. has its value entered in the B.S. column, but on the same horizontal line as the F.S. A new height of collimation is then calculated by addition of the B.S. to the reduce level.

Summary of computation procedures for height of collimation booking

1 *When surveying to establish unknown level values*
(a) Enter known value of T.B.M. in R.L. column.
(b) Read staff held on T.B.M. and enter value in B.S. column.
(c) Add B.S. reading to known T.B.M. value (a + b).
(d) Enter result of (a + b) in height of collimation column.
(e) Read intermediate sights on required locations and enter in I.S. column.
(f) Deduct (unless inverted staff used) I.S. values from height of collimation, and enter result in R.L. column for all readings logged.
(g) Take final reading and enter in F.S. column.
(h) Deduct F.S. reading from height of collimation to give reduce level.

(*Note*: It is recommended that the final reading (g) be taken on a point of known value e.g., a T.B.M., to establish if any closing error is evident, prior to moving the instrument. The check on closing error will prove if re-levelling is required. It is often preferable to compute step (f) in the site office after all readings have been taken.)

2 *When levelling to establish points from given values*
   (a) Enter known value of T.B.M. in R.L. column.
   (b) Referring to the drawings, enter the level values of the points to be levelled in the R.L. column.
   (c) Read staff value when held on T.B.M., and insert in B.S. column.
   (d) Add B.S. reading to T.B.M. value noted in R.L. column.
   (e) Enter result of (c + d) in height of collimation column.
   (f) Deduct reduce level values (b) from height of collimation (e).
   (g) Enter results of (f) in I.S. column.
   (h) Take final reading on separate T.B.M. or point of known value, and enter in F.S. column.
   (i) Calculate value of point read in (h), and compare with known value to ascertain any closing error apparent.

Accuracy checks for height of collimation booking

Partial check Whenever booking levels either by methods 1 or 2 previously explained, checks should be carried out to confirm the accuracy of arithmetical computations.

For height of collimation booking, the basic check is made by finding the separate totals of B.S. and F.S. columns. The Σ F.S. is then subtracted from the Σ B.S. The result of this calculation should equal the difference between the first and the last reduce level values i.e.:

Σ B.S. $- \Sigma$ F.S. = difference in reduce level of start and finish.

Complete check The basic check gives no examination of the intermediate readings which could contain an error, yet go unnoticed. Consequently a more complete check is desirable. This complete check for accuracy of reduce levels, can be made as follows:

   (a) Find Σ B.S. column.
   (b) Find Σ I.S. column.
   (c) Find Σ F.S. column.

(d) Find Σ (height of each collimation × number of applications).

(e) Find Σ reduce levels excluding the first.

(f) Compute Σ B.S. − Σ F.S. columns.

(g) Compute Σ F.S. + Σ I.S. + Σ reduce levels excluding first.

(h) To check: Σ (g) should equal Σ (d).

(i) Find difference in reduce level of start and finish which should equal (Σ B.S. − Σ F.S.).

An illustration of this check is shown in Fig. 3.1.

Rise and fall level book

The rise and fall method may be adopted as an alternative manner of booking and calculation to that of the height of collimation. It is mainly a matter of personal choice regarding the method used. The height of collimation method involves less arithmetic, but errors are not so easily apparent during the course of levelling, especially when working from given reduce level values. The author prefers the height of collimation method, although the rise and fall method is still frequently practised.

An example of rise and fall booking is given in Fig. 3.1, from which it is evident that the height of collimation column used in the alternative height of collimation method, has been replaced by two columns headed 'rise' and 'fall' respectively.

Booking methods (with reference to 1 and 2 on page 61)

1 When levels taken on site are booked, the first reading from the T.B.M. or datum, is entered in the B.S. column. The intermediate levels are then read and entered in the I.S. column. The final reading is taken prior to moving the instrument and the level reading entered in the F.S. column. The rise and fall values, and resulting reduce levels can then be evaluated from the booked information.

2 Before establishing the planned levels on site, and whilst in the site office, enter reduced level values and the rise and fall between each in the level book. After setting up the level and reading a B.S. on the datum, values for staff reading (i.e., I.S. and F.S.), can then be calculated by subtraction or addition of the rise or fall to each. The following summary illustrates methods (1) and (2).

Computation process for rise and fall booking

1 *When surveying to establish unknown level values*
(a) Enter known value of T.B.M. in R.L. column.
(b) Read staff held on T.B.M. and enter in B.S. column.
(c) Read intermediate sights and enter in I.S. column.
(d) Compare first I.S. with B.S. to establish rise or fall.
(e) Add rise or subtract fall from value of T.B.M. reduce level to give R.L. at this point.
(f) Compare next I.S. (or F.S.) with previous I.S. to establish rise or fall.
(g) Add rise or subtract fall from previous reduce level value to give R.L. at this point.
(h) Repeat steps (f) and (g) as required, but take note of any change points during the course of the survey, where a B.S. will have been recorded. In such cases calculations would re-commence from step (d) with the level in a new position.

2 *When levelling to establish points from values given on the plans*
(a) Enter known values of T.B.M. and reduce levels in R.L. column.
(b) Calculate rise and fall between respective R.L. values.
(c) Read staff held on T.B.M. and record reading in B.S. column.
(d) Deduct first rise or add first fall value to B.S. and enter in I.S. as relevant.
(e) Deduct second rise or add second fall value from previous I.S. and enter as next I.S.
(f) Continue as (d) and (e) by addition or subtraction from previous I.S. until F.S. is reached.
(g) Where (as in example on Fig. 3.1) a change point is encountered when a new B.S. is taken, calculations should re-commence from step (d).

Accuracy check for rise and fall method
As with the collimation method, a check on the accuracy of calculations should always be undertaken.

1 Firstly, to show errors quickly, compare (the sum of the backsights minus the sum of the foresights) with (the sum

of the rise column minus the sum of the fall column). If correct the calculations should be equal i.e. Σ B.S. $-$ Σ F.S. = Σ rise $-$ Σ fall columns.

2 If this check proves accurate complete the examination by comparison with the difference between first reduce level and the last reduce level, which should equal Σ B.S. $-$ Σ F.S. columns.

Therefore the whole check becomes:

$$\Sigma \text{ B.S.} - \Sigma \text{ F.S.} = \Sigma \text{ rise} - \Sigma \text{ fall} = \text{first R.L.} - \text{last R.L.}$$

An illustration of this check is shown in Fig. 3.1.

Carrying forward level readings

If a large number of readings is taken from one instrument location, it may be necessary to carry over level booking from one page to the next. Upon meeting the bottom of the page the final I.S. reading is entered in the F.S. column, although it is not actually the last reading of the series. This is done to enable a check of the computations on the completed page. On the new page, the same reading is entered in the B.S. column and a value given to the R.L. column as calculated. The remaining intermediate sights are then taken as normal.

Theodolite booking

Generally the site engineer will have less need than a site surveyor to book theodolite readings. In the vast majority of instances, theodolite work will be required to produce location points from known data, rather than to take readings of existing features.

When field readings are to be taken, the engineer must record these in a logical format on a record sheet. The booking page illustrated in Fig. 3.2 presents a typical example of how the sheet may be designed and compiled. Allowance is made for entry of both horizontal and vertical angles together with space for noting face left, face right and mean angles etc. (The mean angle column is provided to enter the mean value of the face left and face right readings.)

As, during accurate work, more than one round of readings will be taken on each point, a further column provides space for noting the reference point number. It is also useful to allow space for the sketching of a plan diagram.

Theodolite angle booking sheet

Contract HARTOP HOUSING

Survey date 2...6...80   Theodolite set up on station ......O......

Booker T. FIELD

Surveyor R.J. SMITH   Theodolite height....1.35.m....

Operation LOCATION OF STATIONS A/B/C

Horizontal angles

Station point	Face left (F.L.)	Face right (F.R.)	Mean angles	Provisional angles	Horizontal angle	Layout diagram	Notes and remarks
A (R.P.)₁	00°-02'-20"	180° 02' 00"	00°-02'-10"	00°-00'-00"	00°-00'-00"		20" theodolite Nº 3462.11 used with approximation to nearest 5".
A (R.P.)₂	90°-07'-30"	270°-06'-40"	90°-07'-05"	00°-00'-00"			
B₁	25°-35'-10"	205°-34'-40"	25°-34'-55"	25°-32'-45"	25°-32'-50"		
B₂	115°-40'-30"	295°-39'-30"	115°-40'-00"	25°-32'-55"			
C₁	84°-40'-50"	264°-40'-10"	84°-40'-30"	84°-38'-20"	84°-38'-42"		
C₂	174°-46'-30"	354°-45'-50"	174°-46'-10"	84°-39'-05"			

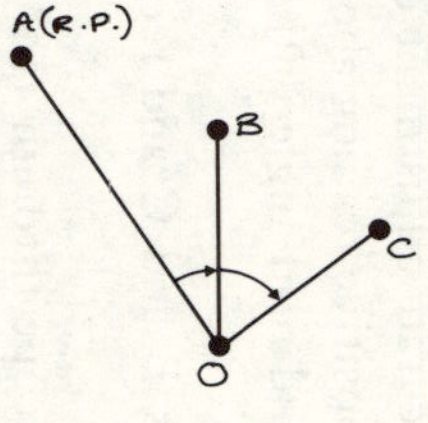

Vertical angles

Point	Target height	Face left (F.L.)	Face right (F.R.)	Reduced (F.L.)	Reduced (F.R.)	Mean	Notes and remarks
A	1.38 m	87°-35'-20"	272°-24'-50"	02°-24'-40"	02°-24'-50"	02°-24'-45"	20" theodolite Nº 3462.11 used.
B	1.38 m	88°-40'-40"	271°-20'-30"	01°-19'-20"	01°-20'-30"	01°-19'-55"	
C	1.38 m	93°-30'-50"	266°-28'-50"	03°-30'-50"	03°-31'-10"	03°-31'-00"	

Fig. 3.2   Theodolite angle booking sheet

Horizontal angles

During site work, the procedure for angle reading will be as follows (with reference to sketch diagram on Fig. 3.2).

  (a) Set up theodolite on point o.

  (b) Choose the reference point (R.P.), usually the most distant point (here taken as point A).

  (c) Set the horizontal scale to between 0°–05′ and 0°–10′ (to allow for accurate adjustment on to the R.P. and still maintain a positive reading above 0°–00′).

  (d) Sight the R.P. and read angle with the theodolite face left (F.L.).

  (e) Turn to stations B and C and record angles as read from the scale.

  (f) Transit the telescope through 180° vertically to make the instrument face right (F.R.) and take a second reading on C.

  (g) With the instrument still F.R. turn back towards points B and A and read and record horizontal angles on each.

  (h) Commence second round of angles by resetting the telescope to F.L. and the horizontal circle to read approximately 90°, then read station A. Repeat steps (d) to (g) noting the readings for the second round.

Horizontal angles are booked as they are taken in face left and face right columns. The mean angle is then calculated by halving the difference between F.L. and F.R. readings for each station point. The mean applies to minutes and seconds only as the F.L. degrees are carried through.

The provisional angle in each round is then calculated by subtracting the mean R.P. reading from other mean readings taken in the same round, i.e. $B_1 - A$ (R.P.)$_1$ = provisional angle B_1, likewise $C_1 - A$ (R.P.)$_1$ = provisional angle C_1.

The actual horizontal angle is then computed by finding the mean of the two provisional angles for each station point. (See Fig. 3.2 for example.)

Vertical angles

Vertical angle readings may be taken and booked following the horizontal angle readings.

Usually a target is set above each station to correspond with the theodolite telescope height. One round of angles is generally taken on both face left and face right of the instrument. Vertical angles are booked in the face left and face right columns during reading on site. Once read, it is generally necessary to reduce each reading – each reduced value being recorded in the reduced F.L. and reduced F.R. columns. The actual vertical angle is obtained by calculating the mean of the reduced F.L. and F.R. angle readings.

Frequently when carrying out large surveys, an assistant books the readings as they are read and called out by the engineer. This reduces the surveying time required and minimizes errors in either reading or logging.

T.B.M. and grid station reference list

Figure 3.3 details a typical 50 m site grid layout. Following the installation of co-ordinated and levelled grid reference points, a reference list compiled to indicate the values of each station is of considerable use to the site team.

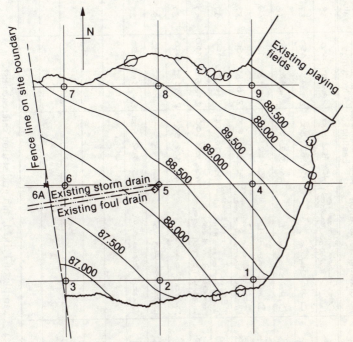

Fig. 3.3 Site grid locations (refer to Fig. 3.4 for grid reference co-ordinates)

T.B.M. and grid station reference list.   Contract: *Natap. ha.sig. estate......*

Station number	Station/level location	Co-ordinate references		T.B.M. level values	Date of establishment	Notes on origin of grid and level survey
		Eastings	Northings			
1	Cross etched in steel pin level on top of pin (encased in conc)	456402·500	246485·450	88·410	6th May 1980	
2	Ditto Station 1	456352·500	246485·450	87·580	"	
3	Ditto Station 1	456302·500	246485·450	86·923	"	
4	Ditto Station 1	456402·500	246535·450	89·443	"	
5	Cross etched on manhole cover level on painted cross area.	456352·500	246535·450	88·355	Existing	* Value given by Architect. Station at 5 etched on to existing manhole cover (provided by client).
6A	T.B.m on end on road 1 (existing) Painted cross on kerb R.H.channel	456294·200	246535·450	87·760	"	* Station 6A Nail in road kerb joint. (Value provided by clients Architect)
6	Cross etched in steel pin level of T.B.m on top of pin.	456302·500	246535·450	87·684	6th May 1980	
7	Ditto Station 6	456302·500	246585·450	88·723	"	
8	Ditto Station 6	456352·500	246585·450	89·411	"	
9	Ditto Station 6	456402·500	246585·450	88·398	"	

Fig. 3·4 Example of T.B.M. and grid station reference list

The example given in Fig. 3.4 shows a suggested layout for the list which may be displayed on the office pinboard and copied into the engineer's site level book. It is unwise to distribute copies of the original list among too many colleagues, as any update of the information (due to re-establishment of a station after damage etc.) may go unamended on a little-used record and may subsequently be incorrectly used. The engineers' level book numbers in which copies have been listed, should be noted on the office chart, these books can then be called in for updating along with the main sheet if alterations are required. Similarly, any stations made obsolete must be struck off the list and out of the field books.

The list shown in Fig. 3.4 provides columns for recording the following:

1 station number
2 location
3 co-ordinate value – eastings/northings
4 T.B.M. level value
5 date of establishment
6 notes on origin of grid.

Sketch plan for grid station setting

When grid stations are being located, a sketch plan noting the bearings and distances used should be drawn up and kept in the site file.

For the site shown on Fig. 3.3, the client provided two co-ordinated locations and values, these being at stations 5 and 6A. From these points, the engineer will have decided where further grid stations were to be located, and calculated the distances and bearings to each. The example shown in Fig. 3.5 indicates that the theodolite is to be set up at existing station 5 and after sighting reference point 6A, bearings are turned and distances measured to locate the new grid stations. The methods used for calculating angle bearings and distances from grid co-ordinates are explained in Chapter 4.

Drainage sight rail record chart

During the process of setting out drainage, sight rails will be erected to establish dig levels. The distance that sight rails are set above the drain inverts will be relative to the existing ground

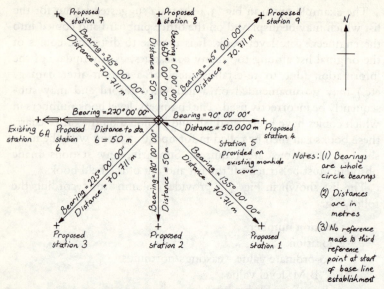

Fig. 3.5 Example of setting out sketch used for location of grid stations on layout shown in Fig. 3.3

levels, and the height needed above ground to enable correct sighting of the traveller. Thus each drain run to be excavated may warrant a change in traveller length, as sight rails cannot always be set at the same height above invert. The engineer should write the level and traveller length on the sight rails. However, as the timber may be frequently re-used throughout the course of the works, such procedures become confusing and less practical, especially if writing is not obliterated before each board is re-set on new profiles. As a safety precaution and record of completed setting out, a drainage sight rail record chart may be formulated for use by other site staff and the site ganger. Such a chart provides easy reference to the recorded information when the engineer is unavailable or elsewhere on the site.

An example of a typical storm drainage layout, giving manhole numbers and cover and invert levels, is shown in Fig. 3.6.

Figure 3.7 gives a compiled example of a drainage sight rail record sheet for the layout shown in Fig. 3.6. Recorded on the sheet are the drain lengths, manhole numbers, the invert levels of each manhole, the specified depth of pipe bed, sight rail levels at each manhole, the computed traveller length, the sight rail decoration colour code, and the date of sight rail erection. The sight rail

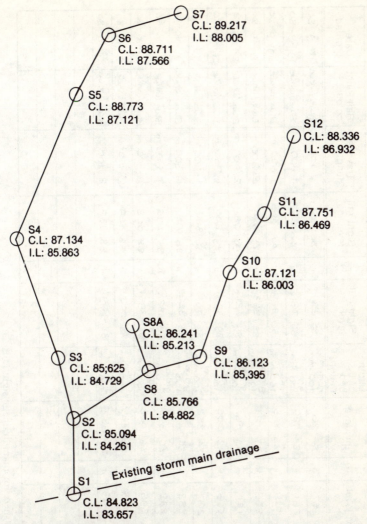

S7
C.L: 89.217
I.L: 88.005

S6
C.L: 88.711
I.L: 87.566

S5
C.L: 88.773
I.L: 87.121

S12
C.L: 88.336
I.L: 86.932

S11
C.L: 87.751
I.L: 86.469

S4
C.L: 87.134
I.L: 85.863

S10
C.L: 87.121
I.L: 86.003

S8A
C.L: 86.241
I.L: 85.213

S9
C.L: 86.123
I.L: 85,395

S3
C.L: 85;625
I.L: 84.729

S8
C.L: 85.766
I.L: 84.882

S2
C.L: 85.094
I.L: 84.261

Existing storm main drainage

S1
C.L: 84.823
I.L: 83.657

Fig. 3.6 Example of typical drainage layout

levels will have been set to suit the site contours and invert levels, recorded in the case of the drain between manholes S4 and S5 as being 2.5 m above invert level. The length of traveller is gauged from the sight rail height above invert plus the depth of drain bed, in this case computed to be 2.5 + 0.1 = 2.600 m. Chart entries should be made at the end of each day's setting out, from the information recorded in the engineer's level book. If error is suspected checking will be easier if records have been compiled.

Record sheet for main drainage sight rail levels

Drainage type Storm.....

M/H No	To	M/H No	Inv. level	To	Inv. level	Depth of bed material	Sight rail level	To	Sight rail level	Traveller length	Rail Paint colour	Date profiled
S 1	"	S 2	83·657	"	84·261	gravel 100 mm	85·657	"	85·261	2·100 m	Blue/white	18-9-80
S 2	"	S 3	84·261	"	84·729	gravel 100 mm	85·761	"	86·229	1·600 m	Blue/white	19-9-80
S 3	"	S 4	84·729	"	85·863	conc. 150 mm	86·229	"	87·363	1·650 m	Blue/white	19-9-80
S 4	"	S 5	85·863	"	87·121	gravel 100 mm	88·363	"	89·621	2·600 m	Blue/white	24-9-80
S 5	"	S 6	87·121	"	87·566	gravel 100 mm	89·621	"	90·066	2·600 m	Blue/white	24-9-80
S 6	"	S 7	87·566	"	88·005	conc. 150 mm	89·366	"	89·805	1·950 m	Blue/white	26-9-80
S 2	"	S 8	84·261	"	84·882	gravel 100 mm	85·761	"	86·382	1·600 m	Blue/white	1-10-80
S 8	"	S 8A	84·882	"	85·213	gravel 100 mm	86·382	"	86·713	1·600 m	Blue/white	1-10-80
S 8	"	S 9	84·882	"	85·395	gravel 100 mm	86·382	"	86·895	1·600 m	Blue/white	1-10-80
S 9	"	S 10	85·395	"	86·003	gravel 100 mm	87·395	"	88·003	2·100 m	Blue/white	3-10-80
S 10	"	S 11	86·003	"	86·469	conc. 150 mm	88·003	"	88·469	2·150 m	Blue/white	3-10-80
S 11	"	S 12	86·469	"	86·932	gravel 100 mm	88·469	"	88·932	2·100 m	Blue/white	3-10-80

Fig. 3·7 Example of record sheet for main drainage sight rails

Record sheet for road excavation profiles

Contract .....Heathfe housing estate.....

Road chainage	Finish road channel levels		Sight rail height above fin: road		Road camber mm	Depth from fin to excavation level	Hardcore depth	Traveller to dig level	Traveller h/c level
	Left channel	Right channel	Left channel	Right channel					
ROAD A									
100·000	91·440	91·440	1·5 m	1·5 m	75 mm	415 mm	250 mm	1·915 m	1·665 m
115·000	91·490	91·490	1·5 m	1·5 m	75 mm	415 mm	250 mm	1·915 m	1·665 m
120·000	91·450	91·450	1·5m / 2·0 m	1·5m / 2·0 m	75 mm	415 mm	250 mm	1·915m / 2·415m	1·665m / 2·165 m
130·000	91·150	91·150	2·0 m	2·0 m	75 mm	415 mm	250 mm	2·415 m	2·165 m
140·000	90·750	90·750	2·0 m	2·0 m	75 mm	415 mm	250 mm	2·415 m	2·165 m
150·000	90·350	90·350	2·0 m	2·0 m	75 mm	415 mm	250 mm	2·415 m	2·165 m
158·000	90·030	90·030	2·0m / 1·5 m	2·0m / 1·5 m	75 mm	415 mm	250 mm	2·415m / 1·915m	2·165m / 1·665 m
165·000	89·750	89·750	1·5 m	1·5 m	75 mm	415 mm	250 mm	1·915m	1·665 m
170·000	89·550	89·550	1·5 m	1·5 m	75 mm	415 mm	250 mm	1·915m	1·665 m
ROAD B									
0·000	90·825	90·825	1·75 m	1·75 m	—	—	—	—	—
5·000	90·950	90·950	1·75 m	1·75 m	—	—	—	—	—
10·000	91·075	91·075	1·75 m	1·75 m	75 mm	275 mm	150 mm	2·025 m	1·875 m
15·000	91·200	91·200	1·75 m	1·75 m	75 mm	275 mm	150 mm	2·025 m	1·875 m
25·000	91·450	91·450	1·75 m	1·75 m	75 mm	275 mm	150 mm	2·025 m	1·875 m
30·000	91·575	91·575	1·75 m	1·75 m	75 mm	275 mm	150 mm	2·025 m	1·875 m
37·259	91·756	91·750	1·75 m	1·75 m	75 mm	275 mm	150 mm	2·025 m	1·875 m

Fig. 3.8 Example of record sheet for road excavation profiles

71

Road profile record sheet

Depending on the existing ground levels and their effect on the road work excavation profiles, various heights of traveller will be used. Generally, the engineer should endeavour to keep each section or length of road at a constant traveller height, but this is not always practical.

A record sheet of road profile sight rail heights and traveller lengths should be kept either on file or pinned on the office wall. Each section of road chainage should be recorded and its appropriate details entered on to the chart. The chart compiled after each stint of setting out, can be referred to by any site personnel. In the event of profile damage to a particular section of road, replacement on site is assisted by reference to the chart.

Fig. 3.8 shows a suitable format for a road profile record sheet.

House block profile record sheet

A similar chart to that devised for drainage and road profile recording should be used to log profile levels erected for use during house construction. As with drainage work etc., the sight rail will be set on site to suit the existing ground contours and final dig levels, thus various profile heights and traveller lengths will be used.

Figure 3.9 shows a typical house block layout for modern terraced housing around a road cul-de-sac. The diagram shows the proposed house finish floor levels (F.F.L.) and crosses to denote the erected profile locations. The profiles may be either on single or double stakes, examples of which are shown in Chapters 5 and 6.

The record sheet shown in Fig. 3.10 provides space for entry of block numbers (or, in the case of other works, the relevant construction areas), finish floor levels, sight rail levels, and traveller lengths to various elements of the construction. Where the record sheet is to be displayed on the office wall, it is more practical to enter as much information on to the chart (i.e., the total of house block numbers and finish floor levels etc.), whilst the chart is still being drawn up.

Drain junction location record sheet

As main drainage is laid, branch junctions are frequently left to take subsidiary drain connections at a later date. Frequently the

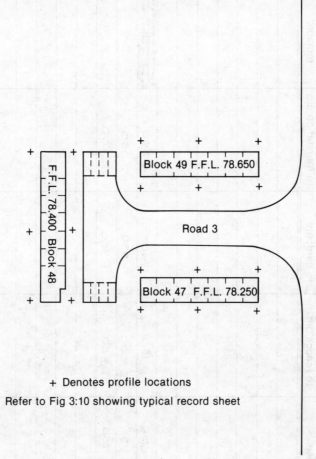

Block 49 F.F.L. 78.650

F.F.L. 78.400 Block 48

Road 3

Block 47 F.F.L. 78.250

+ Denotes profile locations

Refer to Fig 3:10 showing typical record sheet

Fig. 3.9 Typical house block layout showing profile locations for excavation purposes

junctions can only be approximately located to suit pipe lengths, although this will depend on the drain pipe material used. Immediately following drain laying, and prior to backfill, a taped distance of each drain junction location should be made, commencing at the centre of the lowest manhole on a running tape along the drain. The taped lengths are invaluable when used some time after backfill to relocate junctions. If the approximate drain depth is also recorded it will be useful when finding any junctions during re-excavation.

Record sheet for house block reduce dig profiles						Contract: HARTOP HOUSING ESTATE
Block numbers	House block finish floor levels	Sight rail levels	Traveller length to reduce dig	Traveller length to hardcore	Traveller length to foundation dig	Notes and remarks
40	77.500					
41	77.500					
42	76.375					
43	76.500					
44	76.800					
45	76.800					
46	77.450					
47	78.250	79.750	1.750	1.600	2.500	*concrete oversites 100 mm Hardcore oversites 150 mm found 1 m below finished floor level
48	78.400	80.150	2.000	1.850	2.750	'' '' ''
49	78.650	79.850	1.450	1.300	2.200	'' '' ''
50	78.500					
51	78.500					

Fig. 3.10   Example of record sheet for house block reduce dig profiles

Drain junction location record sheet		M.H.No ..6..		To M.H.N.° ..7... Laid on .8./5./8.Q.
Sketch diagram of drain/junctions	Junction number	Distance from MH ..6..	Junction branch size	Left or right hand	Notes

	Junction number	Distance from MH ..6..	Junction branch size	Left or right hand	Notes
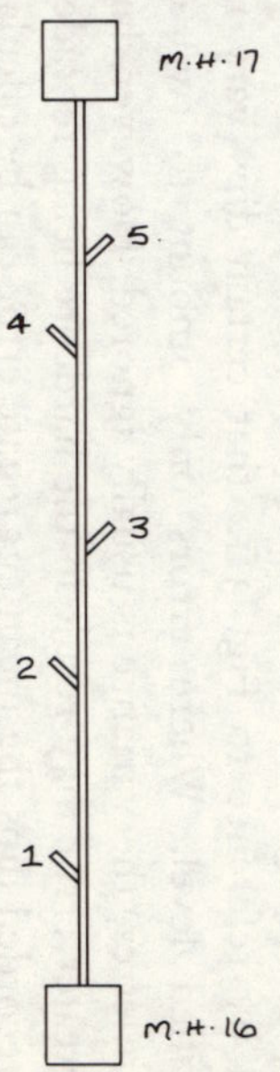	1	5 m	150 mm	Left	Invert level M/H 16 = 82.355 m.
	2	8.62 m	100 mm	Left	Invert level M/H 17 = 82.875 m
	3	10.2 m	150 mm	Right	Main drain length ∅ = 150 mm
	4	18.66 m	150 mm	Left	Approx average drain depth ≅ 1.75 m
	5	19.83 m	100 mm	Right	
	M.H.17	25.7 m	—		
Totals	5 N°	25.7 m	—	—	

Storm. Drainage 150 mm ∅ main.

Fig. 3.11 Example of drain junction location record sheet

75

A drain junction location record sheet such as the example given in Fig. 3.11 should be compiled and filed in the site office. Each sheet can have a sketch of the drain length concerned and columns to record the following:

1 junction number
2 running taped distance from lowest manhole
3 junction branch size
4 the side of the drain (i.e., left or right) where the junction protrudes when viewed from the lowest manhole
5 a notes column to record invert levels and main drain size etc.

Failure to keep accurate records of junction locations will result in considerable losses in time and money, together with an increased possibility of damage to existing drains during excavation.

Road 'dip' level recording

In the course of road construction, hardcore and tarmac levels will require checking. Hardcore levels must be checked prior to tarmac being laid, usually in the presence of a representative from the subcontractors responsible for tarmac laying and also frequently the clerk of works. A similar procedure should also be carried out between the various layers of tarmac to ensure that correct depths of material are used.

During the checking of levels, usually termed 'dips', record sheets can be compiled, with a copy of the agreed levels being given to the other parties. A suggested layout for the sheet in its compiled form is shown in Fig. 3.12. In the example given, dips were taken by taping below a string line held on top of kerb level and pulled taught between the opposing chainage marks. It can be seen by reference to Fig. 3.12 that certain dips vary from the specified level. Where errors only amount to two or three millimetres, the variance is usually ignored. However, large areas of greater error may result in the hardcore being regraded. From the recorded dips the mean aggregate error can be calculated, and claims by the subcontractor for extra material can be agreed or disputed. On areas of wide road where the use of a string line to establish dips would be impractical, the engineer's level will have to be used, the string line being used merely to show the line between kerb side chainage points along which levels are to be taken. Taped distances out from one kerb line can be used to mark

Record sheet for road dip levels

Contract ...Seaview Housing Estate......... Date ..15-9-80.

Dips taken for ...Hardcore....levels........

Road number	Road chainage	Left channel		Centre line		Right channel		Remarks
		Dip taken	Dip specified	Dip taken	Dip specified	Dip taken	Dip specified	
3	CH 0.000 m	Not taken	—	Take later with level	—	Not taken	—	Junction ℄ of road 3 with road 2
3	CH 5.000 m	" "	—	"	—	"	—	
3	CH 8.650 m	233 mm	235 mm	170 mm	160 mm	231 mm	235 mm	Commencement of straight from Bellmouth
3	CH 10.000 m	230 mm	"	166 mm	"	229 mm	"	
3	CH 15.000 m	238 mm	"	158 mm	"	234 mm	"	
3	CH 20.000 m	235 mm	"	155 mm	"	241 mm	"	
3	CH 25.000 m	234 mm	"	157 mm	"	237 mm	"	
3	CH 27.835 m	240 mm	"	160 mm	"	234 mm	"	Crest of vertical curve
3	CH 30.000 m	233 mm	"	170 mm	"	235 mm	"	
3	CH 37.350 m	228 mm	"	171 mm	"	227 mm	"	End road 3
4	CH 10.000 m	229 mm	235 mm	171 mm	160 mm	230 mm	235 mm	Commencement of straight from Bellmouth
4	CH 15.000 m	230 mm	"	165 mm	"	235 mm	"	Start left hand curve
4	CH 18.500 m	235 mm	"	168 mm	"	227 mm	"	
4	CH 20.000 m	226 mm	"	159 mm	"	230 mm	"	
4	CH 26.350 m	241 mm	"	171 mm	"	236 mm	"	End left hand curve
4	CH 35.000 m	237 mm	"	163 mm	"	228 mm	"	End road 4

NOTE: ① Dips taken from string line held taught between drainage marks on top of kerb lines
② Road formation:
  tarmac 20 mm wearing course
  40 mm base course
  75 mm Road base
③ kerb levels 100 mm above finish road
④ roads 3 and 4 = 5.5 m wide with 75 mm camber and both channel lines level

Fig. 3.12 Example of a record sheet for road dip levels

77

the road centre line, or alternatively a pre-cut gauge rod may be adopted and held against the kerb face to denote half the road-width. On wide roads it is often necessary to take dips in both channel lines and at three or more points across the road span. Obviously, with such roads an extended dip record sheet should be drawn up to include extra columns covering the number of additional level points required.

The record sheets can be marked with the road number, chainage lengths and the specified dip levels entered, prior to the road check. The sheets are used on site, and entries of dips made at the time of their measurement. A site file must be used to store the sheets once agreement has been reached between all parties that the recorded information is acceptable.

Library of calculator programs

In Chapter 2 and elsewhere within the book, examples of programs for use with programmable calculators are shown.

Once a program has been devised, it should be recorded and a library of programs compiled to cover all sections of the engineer's operations. It may be prudent to design a program record sheet to suit the calculator type being used, which may differ from the format presented in Fig. 3.13.

Figure 3.13 illustrates a program record sheet developed by the author for use with the 'Texas TI57' calculator, although the instrument is provided with its own supply of manufacturer's designed record sheets which may be preferred by some operators. As a guide, a record sheet should allow space to insert the following information:

1 title
2 program compiler's name
3 date
4 formulae used
5 program description and notes
6 program step number
7 column for indication of button used
8 calculator key code
9 calculator display
10 area for recording the information stored in memories.

The compiled and completed record sheets should be filed in a stout binder, with copies taken for everyday use, and duplicate

Program record sheet for Texas TI 57 calculator

Program title:

Program drawn up by:	Date:

Formulae used:

Program description and notes

Step number	Press	Key code	Calculator display	Step number	Press	Key code	Calculator display

Information required in memories fed into:
STO 0 −
STO 1 −
STO 2 −
STO 3 −
STO 4 −
STO 5 −
STO 6 −
STO 7 −

Fig. 3.13   Example of pro-forma for program record sheet for Texas
TI57 calculator

Contract drawing register

Site: Hartop housing estate   Work section: Roadworks

Drawing number	Drawing title	Issue status	Revision number/letter and date of issue				
			a	b	c	d	e
A736/01			a	b	c	d	e
,, /02			a	b	c	d	e
,, /03	Road 1 long sections	14-1-80	a	b	c	d	e
,, /04	,, 2 ,, ,,	14-1-80	a	b	c	d	e
,, /05	,, 3 ,, ,,	—	a	14-1-80 b	c	d	e
,, /06	,, 's 4/5 ,, ,,	—	a	14-1-80 b	c	d	e
,, /07	,, 6 ,, ,,		a	b	c	d	e
,, /08	,, 7 ,, ,,		a	b	c	d	e
,, /09	,, 7A ,, ,,		14-1-80 a	b	c	d	e
,, /10	,, 8 ,, ,,	—	a	14-1-80 b	c	d	e
,, /11			a	b	c	d	e
,, /12			a	b	c	d	e
,, /13			a	b	c	d	e
,, /14			a	b	c	d	e
,, /15	Road 1 setting out plan	10-1-80	a	b	c	d	e
,, /16	,, 2 ,, ,,	10-1-80	a	— b	b 10-1-80 c	d	e
,, /17	,, 3 ,, ,,	10-1-80	a	b	c	d	e
,, /18	,, 4/5 ,, ,,	—	a	10-1-80 b	c	d	e
,, /19	,, 6 ,, ,,	—	a	10-1-80 b	c	d	e
,, /20	,, 7 ,, ,,	10-1-80	a	b	c	d	e
,, /21	,, 7A ,, ,,	14-1-80	a	b	c	d	e
,, /22	,, 8 ,, ,,	14-1-80	a	b	c	d	e
,, /23			a	b	c	d	e
,, /24			a	b	c	d	e
,, /25			a	b	c	d	e

Fig. 3.14 Example of contract drawing register

80

sets being kept at head office or in the engineer's own possession. The file can be split to cover programs for various work sections e.g., co-ordinates, roadworks, drainage etc., and indexed to give easy reference.

Drawing register

Throughout the course of a project, it is inevitable that certain working drawings will be revised to accommodate variations in the original design. A common oversight and cause of error, is the failure to discard out-of-date drawings and keep accurate records and files of the latest revisions.

At the commencement of the project, the engineer will be issued with all drawings and construction details appertaining to his section of the works. Before the drawings are filed in drawers or racks, and prior to their use, a drawing register should be compiled listing the drawing number, date of issue and title etc. The register forms a record of those drawings issued and available, and provides a reference index to locate drawings. As drawings are revised, their revision number is shaded in on the register and the date of issue entered.

A typical example of a drawing register page is shown in Fig. 3.14 partly compiled to indicate the method of entry. If the register is kept up-to-date and drawings properly stored in good condition, another source of possible error and time loss is reduced.

4

Commencing on site

Before commencing on site, the engineer will require time to study the plans, Bills of Quantities, and site programme. He should preferably be given the opportunity to spend one or two weeks in the head office, together with the contracts manager and site agent, during which time the planned method of undertaking the project can be agreed. Within this period, initial setting out calculations can be worked and any apparent drawing errors discussed with the architect. The engineer may be responsible for ordering and scheduling materials required for sections of the work under his control. Items such as road kerbs, drain pipes and manhole equipment required early in the construction can be requisitioned.

Once the basis of the project has been evaluated, the initial operations on the site must be considered. The principle tasks such as a site inspection, check on the T.B.M. level and also location of the setting out reference stations, existing drains and site features will be the first priorities. Where setting out stations and levels have been positioned by the architect, the engineer must satisfy himself that these are accurate. The T.B.M., if related to the Ordnance Survey, should be checked against the nearest O.B.M. If no survey stations have been provided, the engineer may have either (a) to agree with the architect the setting out base lines for the works, and formulate his own grid stations from these, or (b) to construct a grid in relation to the National Grid.

The remainder of this chapter will explain the sequence and method of undertaking the initial operations on site.

Site inspection

Before any construction operations are begun, and preferably shortly prior to the contractor being given possession of the site, an inspection of the site's condition should take place with the architect and clerk of works present. The engineer and site agent should represent the contractor and agree with the architect the record of any items of damaged material within the site boundary that could otherwise result in a claim for damages against the contractor at a later date. Such inspections often reveal cracked manhole covers, broken road kerbs, damaged trees (some of which the architect may have specified to be preserved) and broken boundary fencing and walls etc. A copy of the records should be distributed to all parties concerned.

Location of the grid stations

After the site inspection the site grid should be formulated – where the size of the project dictates that this is required. The grid is generally set out from the base lines of stations provided by the architect. The location of existing stations or base lines will be shown on the site plans or alternatively agreed with the architect in the field.

If no survey stations have been installed, the engineer may have to construct his own site grid from local O.S. triangulation stations on the National Grid or the agreed baseline. On smaller projects the architect often details the baseline as the edge of an existing road or building. Where this is less well defined, such as a hedge or fence line etc., agreement with the architect about the actual line taken should be made on site.

Where survey stations have been provided, these will most likely have co-ordinate values relative to the National Grid system. Such stations are frequently provided on larger projects, and may be set at random locations in the field. A minimum of three reference stations should be given, but on larger sites more may be necessary. It is also advantageous to have a co-ordinated survey point upon which reference can be made from any two survey stations, on for example, a clock tower or church spire visible from the site. It is useful for sighting and field work purposes if extra stations are set out in addition to those originally provided and it is helpful if these are located to correspond with the grid square structure of the site. Such grids are usually in the form of 50 or 100 m squares, although lesser or greater distances may be used if desired.

Setting stations to correspond with grid squares has the following advantages:

1 easier calculation as the grid stations have more regular co-ordinates
2 bearings are easier to establish as grid lines run north/south and east/west i.e., 0°, 90°, 180°, and 360° bearings
3 reinstatement of a damaged station is less tedious

Against these benefits, the system has the following disadvantages:

1 the grid stations may not suit the ground contours for sighting purposes
2 during the initial stages of construction, the stations may sit in the line of road, drainage or foundation excavation etc., and be quickly displaced.

Generally, the method of adopting a grid system rather than placing stations at arbitrary locations will depend on the site lay-out. Stations given co-ordinate values located in safe random positions, may prove more durable but will result in more complex calculations for bearings and more tedious setting of theodolite scales to correspond with the angles concerned.

Once the location of the survey stations has been decided, their installation may involve the use of co-ordinate calculation and setting out techniques. Co-ordinate calculations are used to determine the bearings and distances between stations and co-ordinated setting out points. The initial references will be taken from those values provided on the survey stations given or alternatively the nearest O.S. reference stations.

Co-ordinates

During recent years the use of co-ordinate references for setting out has been extended from large civil engineering projects and roadwork schemes to use on general construction sites, especially those covering a large area. The use of computers in design work has enabled architects and designers to detail the setting out points with a high degree of accuracy.

If we consider as an example a medium sized housing project, it is now commonplace for the drawings to show co-ordinated setting out details for road centre lines, house locations, drain manhole locations and grid squares. Alternatively, grid station and road line co-ordinates may be given, and houses and drains are then set out from these by traditional methods. Whichever scheme is used, at some stage calculation of co-ordinates will be necessary.

84

National Grid

The National Grid which covers the whole of England and Scotland, divides the area in a series of 100 km square grid lines running north and east from a false origin lying to the south-west of the British Isles.

Values of the grid are usually given on the site layout drawings (often plans reduce the grid to 100 m squares), but can otherwise be obtained from O.S. maps. Whole circle bearings are taken from the north line of the National Grid known as 'grid north'. Bearings between stations relate to this 'grid north' origin.

As the National Grid originates from a map projection and represents the curved surface of the earth on a plane, points on the grid are altered slightly accounting for the curvature of the earth. Due to this, distances and bearings calculated from the grid will not always agree with the equivalent field measurement. Where necessary, this variance will have to be adjusted by use of a scale factor which varies throughout the country. Usually only the distances are adjusted, as angle bearings of less than 10 km in length are assumed to agree with those calculated. The approximate scale factor value can be obtained by reference to a table and by using the following calculations.

Scale factor calculation

The distances stated on the Ordnance Survey stations based on the National Grid are projected distances, and differ from physical distances (i.e., actual ground distances measured). To convert projection distances to ground distances it is necessary to use a local scale factor (adapted from CIRIA, 1976).

$$\text{Thus the ground distance} = \frac{\text{projection distance}}{\text{local scale factor}}$$

$$\therefore \text{local scale factor} = \frac{\text{projection distance}}{\text{ground distance}}$$

Table on page 86 gives the approximate value of the local scale factor for any part of the country.

Co-ordinate terminology

When working with co-ordinates, values for the distance east and north of the National Grid origin will be given. Eastings are al-

Table 4.1 Scale factor calculation table

National Grid Eastings (km)		Scale factor	National Grid Eastings (km)		Scale factor
400	400	0.999 60	210	590	1.000 04
390	410	60	200	600	09
380	420	61	190	610	14
370	430	61	180	620	20
360	440	62	170	630	25
350	450	63	160	640	31
340	460	65	150	650	37
330	470	66	140	660	43
320	480	68	130	670	1.000 50
310	490	70			
300	500	72			
290	510	75			
280	520	78			
270	530	81			
260	540	84			
250	550	88			
240	560	92			
230	570	0.999 96			
220	580	1.000 00			

ways quoted first. From these values quadrant bearings are calculated for the line between stations, relative to the north/south meridian and the quadrant of the whole circle the line bisects. By addition or subtraction, depending on the quadrant bearing's location within the whole circle, the whole circle bearing (W.C.B.) can be calculated.

These points are further explained as follows. The difference between the east values of two co-ordinates is known as the partial eastings and the difference between two north values are known as the partial northings, both abbreviated by using a triangular symbol to $\triangle E$ and $\triangle N$.

Quadrant bearings

Quadrant bearings are angles measured to the east or west of the north/south meridian.

The angle when calculated will bisect one of the four quadrants of a whole circle. The quadrant in which the bearing lies depends on the value of the partial eastings ($\triangle E$) and partial northings ($\triangle N$).

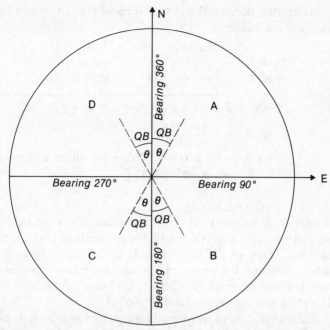

Fig. 4.1 Diagram showing location of quadrant bearings before conversion to whole circle bearings

With reference to Fig. 4.1, the following can be seen to apply:
Where quadrant bearing lies in:

Section A the bearing is termed N.θ°.E.
Section B the bearing is termed S.θ°.E.
Section C the bearing is termed S.θ°.W.
Section D the bearing is termed N.θ°.W.

For example: If the value for θ in Section A was 45°, this would be termed N. 45°. E.

Quadrant bearings are useful when presented in this manner, as the angles given (i.e., 0°–90°) are easily used with mathematical tables when ascertaining tangents, sines and cosines etc. However, with electronic calculators capable of quickly finding trigonometrical functions of angles over 90°, this advantage is less important.

The quadrant bearing is calculated by dividing the partial eastings by the partial northings, the result of which gives the tangent of the bearing, which is then converted into degrees, minutes and seconds. That is, $\tan QB = \frac{\Delta E}{\Delta N}$. The worked example following shows the procedure involved.

Using two stations B and C shown in Fig. 4.6 the co-ordinate values are as follows:

	Eastings	Northings
Point B	473 625.641	262 457.192
Point C	473 661.824	262 442.316

$$\therefore \triangle E = + 36.183 \quad \triangle N = -14.876$$

$$\therefore tan\ \theta = \frac{+36.183}{-14.876}$$

$$\therefore tan\ \theta = -2.432 \quad (\textit{Note: the minus sign is ignored})$$

$$\therefore \theta = S.\ 67°\ 39'\ 03''\ E.$$

Whole circle bearings

Whole circle bearings offer the advantage of instant field use, as no further correction is required for the angle to correspond with the 360° scales of the theodolite. It is also easier to ascertain in which direction the projected line will lay relative to grid north. Whole circle bearings (W.C.B.s) are obtained from grid north 0° in a clockwise direction round to 360°.

The conversion of quadrant bearings into whole circle bearings can be simplified by remembering the layout shown in Fig. 4.2 and by reference to the values of $\triangle E$ and $\triangle N$.

For example: Where $\triangle E$ are positive and $\triangle N$ negative, the bearing must fall into the S.E. quadrant B, with the result that the quadrant bearing must be deducted from 180° to give the W.C.B.

Similar assumptions can be applied to all quadrants of the circle as follows:

1 When the quadrant bearing (Q.B.) is in the NE quadrant i.e., 0° to 90° (+E and +N), no adjustment to the angle is required to give the W.C.B.
2 When the Q.B. is in the SE quadrant i.e., 90° to 180° (+E and −N) the angle must be subtracted from 180° to give the W.C.B.
3 When the Q.B. is in the SW quadrant i.e., 180° to 270° (−E and −N) the angle must be added to 180° to give the W.C.B.
4 When the Q.B. is in the NW quadrant i.e., 270° to 360° (−E and +N) the angle must be subtracted from 360° to give the W.C.B.

The worked example following shows the procedure involved. The W.C.B. for the line for which the quadrant bearing was

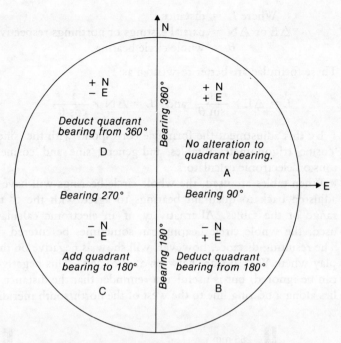

Fig. 4.2 Diagram showing relationship of quadrant bearing when converting to whole circle bearings

calculated between stations B and C in Fig. 4.6 can be obtained as follows:

Given that the quadrant bearing B/C = 67° 39′ 03″, $\triangle$E are positive and $\triangle$N are negative.

∴ Quadrant bearing is in SE quadrant, and thus deducted from 180° to find the W.C.B.

∴ W.C.B. = 180° – 67° 39′ 03″
= 112° 20′ 57″

Note: Using a programmable calculator, a program can be devised and used to calculate the W.C.B. direct from the partial eastings and northings.

Calculation of distance between co-ordinated points Once whole circle bearings have been obtained, distances between co-ordinated points may be calculated using the formula, $L = \triangle E \times$ cosecant θ or $L = \triangle N \times$ secant θ, according to whether $\triangle E$ or $\triangle N$ is the larger.

Where L = distance
$\triangle E$ or $\triangle N$ = partial eastings or northings respectively
θ = whole circle bearing.

These formulae are better re-written as:

$$L = \triangle E \times \frac{1}{\sin \theta} \quad \text{and} \quad L = \triangle N \times \frac{1}{\cos \theta}$$

By this adjustment the formulae correspond with the 'sine' and 'cosine' trigonometrical tables, and general 'sine' and 'cosine' buttons on electronic calculators.

When tables are used, the whole circle bearings will have to be adjusted back to quadrant bearings to agree with the 0° to 90° range of the tables. Alternatively, if an electronic calculator is used, the whole circle bearing can sometimes be entered direct. The resulting distance, however, will show as negative on the display where W.C.B.s are between 180° to 360°. This negative sign can be ignored, but is useful as a reminder that the distance given lies along a bearing line to the west of the north/south meridian.

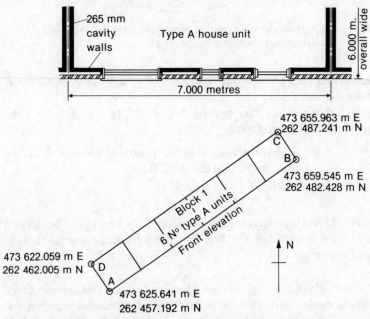

Fig. 4.3 Example of terraced house block and house unit using co-ordinates to plot location (scale factor adjustment not being required)

It has been explained on page 85 that when co-ordinates derived from the National Grid are used, adjustment of the calculated distances by a local scale factor may be necessary to obtain the actual physical ground distance. In practice it will depend upon the wishes of the architect whether the scale factor should be applied in every instance. It is frequently applied when undertaking roadworks designed from National Grid co-ordinates, but not always in drainage and structures work.

An example of where use of the scale factor could lead to difficulties can be demonstrated using a terrace block of houses detailed on the site plans shown in Fig. 4.3.

The individual house plans show each house as 7 m wide from centre to centre of party wall cavities. The terrace block is made up of six identical houses. The overall length of the block is, therefore, 7 m × 6 plus the addition of half an external wall at each end. The external walls are 265 mm wide, thus each half wall is 132.5 mm, this being the addition at each end of the block. Therefore, from the house plans, the total length of the block must be 7 × 6 + (2 × 0.1 325 m) = 42.265 m.

The site setting out plan gives co-ordinate values for each corner of the terrace block. Taking the two front co-ordinates at each end, the distance between can be calculated as follows.

(a) Calculate the bearing between co-ordinates

$$\text{tan quadrant bearing} = \frac{\Delta E}{\Delta N}$$

$$\therefore \tan = \frac{+\,33.904}{+\,25.236}$$

$$\therefore \tan = 1.343\ 477\ 6$$

$$\therefore \text{quadrant bearing} = \text{N. } 53°\ 20'\ 18''\ \text{E.}$$

(b) Calculate the W.C.B.

As the quadrant bearing is in the NE quadrant no adjustment is necessary to obtain the W.C.B., as this will be identical in this instance.

(c) Calculate the distance

$$L = \Delta E \times \frac{1}{\sin \theta}$$

$$\therefore L = +33.904 \times \frac{1}{\sin 53°\ 20'\ 18''}$$

$$\therefore L = +33.904 \times 1.2466$$

$$\therefore L = 42.265 \text{ m}$$

It can be seen from the distance obtained that the co-ordinates are set to correspond with the individual house plans, and if a scale factor were applied, the block length could be incorrect. Obviously the block length should correspond with the plan length to achieve the house sizes required. In such cases the architect's decision should be requested, to agree the method used.

Computing co-ordinate values On occasions, it will be necessary to calculate the co-ordinate values of a point of known bearing and distance from the co-ordinates of an existing station.

The calculation is merely an addition or subtraction to the eastings and northings from the existing station co-ordinates. When using an electronic calculator, whole circle bearings may be used direct and the result will show a positive or negative value indicating whether the co-ordinates are larger or smaller. The formulae are as follows:

difference in eastings = bearing distance $L \times$ sine of whole circle bearing θ

$$\therefore = L \times \sin \theta$$

difference in northings = $L \times \cos \theta$

Worked example: (with reference to the house 'Block 1' in Fig. 4.3)

The co-ordinates of A are given as 473 625.641 m E and 262 457.192 m N. The co-ordinates of point B could be calculated from point A as follows:

Given that the whole circle bearing of line AB is known to be 53° 20′ 18″ and the distance AB is known to be 42.265 m.

Difference in eastings = 42.265 × sin 53° 20° 18″
= 42.265 × 0.802 175 3
= +33.904 m

Difference in northings = 42.265 × cos 53° 20′ 18″
= 42.265 × 0.597 088 6
= +25.236 m

∴ Co-ordinate values for B are:

eastings = 473 625.641 m + 33.904 m
= 473 659.545 m E.

northings = 262 457.192 m + 25.236 m
= 262 482.428 m N.

It can be seen from the known values of 'point B' in Fig. 4.3 that these are the correct co-ordinates.

92

Program record sheet for Texas TI 57 calculator

Program title: CALCULATION OF QUADRANT BEARING AND BEARING LINE DISTANCE FROM CO-ORDINATE VALUES.

Program drawn up by: R.W. MURPHY Date: 16 - 7 - 80

Formulae used: $\dfrac{\Delta E}{\Delta N} = \tan$ quadrant bearing, $\text{LENGTH} = \dfrac{\Delta E \times \frac{1}{\sin} \text{ Quadrant bearing}}{\Delta N \times \frac{1}{\cos} \text{ Quadrant bearing}}$ or

Program description and notes

The program calculates the quadrant bearing and object distance when information as follows is entered into store memories:—

Evaluate partial eastings (ΔE) and northings (ΔN) and store as stated below. Depending if ΔE is larger than ΔN or vice-versa the larger value should be used to obtain the distance calculation. Therefore where eastings (ΔE) are greater use GTO 1 to obtain length after finding angle or GTO 2 if northings (ΔN) value is greater

<u>Results</u> Quadrant bearing can be read direct from the calculation display in degrees, minutes and seconds. The length will be displayed in metres

Step number	Press	Key code	Calculator display	Step number	Press	Key code	Calculator display
1	LRN	—	00-00	26	=	85	25-00
2	RCL 1	33-1	01-00	27	R/S	81	26-00
3	÷	45	02-00	28	LRN	—	0
4	RCL 2	33-2	03-00	29	RST	71	0
5	=	85	04-00				
6	2nd Inv tan	-20	05-00				
7	2nd Inv D.M.S.	-26	06-00				
8	STO 3	32-3	07-00				
9	R/S	81	08-00				
10	2nd LBL 1	86-1	09-00				
11	RCL 3	33-3	10-00				
12	2nd D.M.S.	26	11-00				
13	2nd Sin	28	12-00				
14	1/x	25	13-00				
15	×	55	14-00				
16	RCL 1	33-1	15-00				
17	=	85	16-00				
18	R/S	81	17-00				
19	2nd LBL 2	86-2	18-00				
20	RCL 3	33-3	19-00				
21	2nd D.M.S.	26	20-00				
22	2nd COS	29	21-00				
23	1/x	25	22-00				
24	×	55	23-00				
25	RCL 2	33-2	24-00				

Information required in memories fed into:
STO 0 – /
STO 1 – Enter ΔE value in metres
STO 2 – Enter ΔN value in metres
STO 3 – /
STO 4 – /
STO 5 – /
STO 6 – /
STO 7 – /

Fig. 4.4 Calculator program to compute quadrant bearing and bearing distance value from co-ordinates

Standard sheet for co-ordination calculation and recording

Instrument set on stn. ref.	Eastings value	Northings value	ΔE	ΔN	Site sta. ref.	Quadrant bearing	Locate sta. ref.	W.C.B. to locate	Plan distance	Distance after scale factor Adj.
B	473 625·641	262 457·192								
	473 657·666	262 486·281	+32·025	+29·089	A	N 47° 45' 01·63"E	A	47° 45' 01·63"	43·264 m	
	473 661·824	262 442·316	+36·183	−14·876	−	S 67° 39' 03·07"E	C	112° 20' 56·9"	39·122 m	

Fig. 4·5   Example of standard sheet for co-ordination calculation and recording

Programmable calculator use in co-ordinate calculations

A programmable calculator is excellent for co-ordinate calculations. Programs can be drawn up to assist computation such as the example shown in Fig. 4.4 (for use on a Texas TI57 calculator). This program will calculate the quadrant bearing and distance between co-ordinates once the partial eastings and northings have been stored in the relevant memories. For the purposes of this program, partial eastings and northings are always regarded as positive.

Recording of co-ordinate calculations and worked example

It is important that accurate records are kept of co-ordinate calculations. A pro-forma sheet should be drawn up, such as the example in Fig. 4.5. During calculations, the values obtained should be entered on to the sheet direct from the calculator display or as read from trigonometrical tables. The worked example in Fig. 4.5 has been compiled from the co-ordinate calculation example in Fig. 4.6.

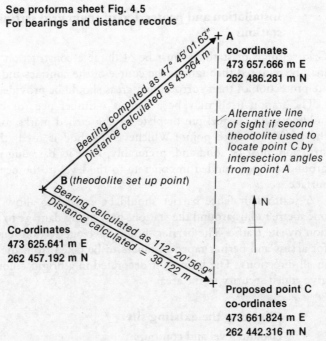

See proforma sheet Fig. 4.5
For bearings and distance records

Bearing computed as 47° 45′01.63″
Distance calculated as 43.264 m

A
co-ordinates
473 657.666 m E
262 486.281 m N

Alternative line of sight if second theodolite used to locate point C by intersection angles from point A

B *(theodolite set up point)*

Bearing calculated as 112° 20′56.9″
Distance calculated = 39.122 m

Co-ordinates
473 625.641 m E
262 457.192 m N

N

Proposed point C
co-ordinates
473 661.824 m E
262 442.316 m N

Fig. 4.6 Example of co-ordinated station layout from two existing stations A and B to proposed station located at point C

With reference to the information given previously in this chapter, the example in Fig. 4.6 requires that point C be set out from a theodolite erected at point B.

Summary of procedure:
(a) Work out the whole circle bearing from B to A.
(b) Compute the whole circle bearing from B to C.
(c) Calculate the distance from B to C.
(d) Having calculated and recorded the bearings B/A and B/C and distance B/C, set the theodolite on point B, sight station A with the horizontal scale pre-set to the required bearing, turn to the calculated bearing of point C and measure the distance.

Note: This method is explained using the procedure of one theodolite and measured distances. As an alternative, two theodolites could be used, the second instrument being set on point A and bearings A/C and B/C used. Point C will then be positioned where lines of sight from both instruments intersect.

Installation and protection of main grid setting out stations

Grid setting out stations must be of durable construction as they may be required throughout the course of the contract and adequate protection of their surrounding areas should be provided.

The station itself may be a nail in a timber peg, or concrete holding secure a steel pin or plate having etched marks to denote the actual reference point. Whichever method is used, the construction must be solid and, preferably, have its base dug into the ground and surrounded in concrete to the level of the peg or pin surface.

A scaffold or fence barrier should be erected to allow at least one metre radius around the station, preventing damage to the station by site traffic. The barrier should have one section removable for access and permit taped distances to be taken from the station in all directions. The barrier, if decorated in a bright colour, will give added security to the area.

Checking the existing site

Ground level and contours

Before any topsoil and reduce dig excavation is commenced, the existing ground levels must be checked and compared against the architect's site survey drawings.

Where drawings show existing levels on a grid pattern e.g., 10 m squares covering the site, the check should be made on corresponding locations. If only contour lines are detailed, then the site should be gridded at say 10 m intervals and the recorded levels interpolated to check if the contour lines correspond.

The level grid can be reconstructed on site by the use of range rods lined and measured between station pegs previously set at the grid locations – various areas being covered during each day's surveying.

Figure 4.7 shows a typical site level grid covering a small area of site. Levels have been taken every 10 m between grid stations. Contour lines are shown interpolated from the recorded levels. When levels are taken, each grid line must be referenced and correct booking is essential. Readings taken on site should be carefully logged, relative to the height of collimation for each section of levelling, the reduced levels being calculated in the site office at the end of each day's surveying. Upon completion of the survey, a plan should be drawn up on tracing paper to a corresponding scale of that used on the architect's site survey drawing. Once all levels have been taken and recorded on the plan, it can be overlaid on the issued survey drawing so that the grid stations correspond. Both plans showing sets of levels or contours can be compared with the detail on the issued survey drawing, which is visible through the plan recorded on tracing paper.

The survey can either then be agreed or disputed, with the architect and any areas of large discrepancy re-checked before earthmoving begins.

Existing features

Following the check on existing ground levels, existing works such as manhole locations and invert levels, road access spurs, buildings and boundaries should be checked.

From the grid setting out stations of known location, angles and distances can be taken to all such existing features and checks made to ensure their location corresponds with the site plans. Any areas of variation between the drawn details should be recorded and passed on to the architect for assessment and re-design if required.

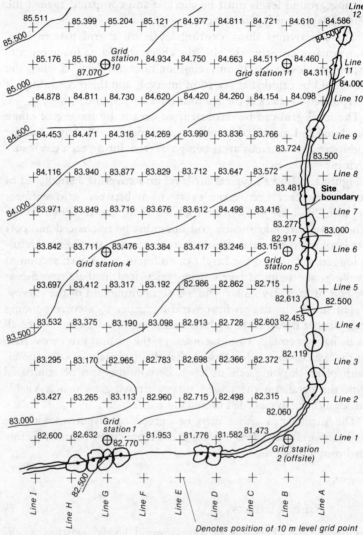

Fig. 4.7 Typical site grid and contour layout over a section of site

Locating existing services

It is highly probable that existing service installations will be present on some areas of a site, especially those in urban districts. The location of these may not always be denoted on the architect's plans and, even when they are, the position scaled from the plans may not agree with the precise site location. It is frequently the responsibility of the contractor to locate existing services e.g., gas, water, electricity, telephone etc., and failure to do so accurately inevitably leads to their damage at some stage of the works. Not only is it costly to pay for the repair of damaged services, but safety aspects must also be considered.

Statutory authorities are most co-operative in supplying details from their records of existing service routes. Some will also provide a cable or pipe detection service where records are dubious, and actually peg the service route on site. The engineer should therefore contact all the statutory authorities with regard to accurate location of existing supplies.

Any overhead electricity cables or telephone lines must be guarded by the erection of gantries and the safety regulations regarding work in the proximity of these adhered to. Buried high pressure mains of gas and water are often identified by concrete markers, and should be fenced off by the contractor – allowance for the cost of this having been covered when pricing the job. Where such mains have to be crossed the correct protection and safe working methods must be used.

Initial setting out

Once the site grid and existing features have been checked and proven, excavation of the topsoil is usually the first operation on a virgin site.

Whilst the site checks are being carried out, the site compound and office accommodation may well be under construction. The area occupied by these temporary works frequently requires careful consideration and siting, ensuring that no area of the new works will be infringed and that connection for temporary services is available at the most economic cost.

The storage area for topsoil (if this is to be retained on site) also requires careful setting out. The site chosen for topsoil storage must obviously be clear of areas of construction. Means of access around the construction areas must be maintained and yet the location must allow adequate access at a later date when topsoil is to be respread over the landscaped areas. The perimeters of the topsoil storage areas should be marked by tall stakes at each corner, within which the topsoil must be contained. The architect may have stipulated a maximum height for the heaps – the storage areas required can therefore be calculated using the measured amounts of topsoil excavation in the Bill of Quantities.

Initial calculations

Whilst the topsoil is being excavated, calculations for reduced dig levels, road profiles and drainage works may be commenced in accordance with the procedures advised in the relevant chapters.

Whenever possible, rough setting out sketches copied from the plans should be used and taken on site. Preliminary work on calculations and sketches must be completed during the early period of the job if possible, as less time will be available once the project is in full swing. Wet days, when work outdoors is limited, should not be wasted – much preparatory work can be done in the site office.

Delegation of duties

Depending upon the size of the contract and the amount of setting out required, more than one site engineer may be employed. Where a number of engineers are engaged, it is usual for grades of superiority ranging from senior to junior to be delegated. The senior (or 'chief' as he may be termed) engineer is generally responsible to the site agent, or alternatively, if a large amount of engineering works is required, direct to the contracts manager.

The sequence of weekly operations will be dictated by the overall site works programme. The detailed daily tasks will be amplified at weekly site meetings. From these agreed planned objectives, the senior engineer will delegate sections of the work to his subordinate staff.

Middle and junior grade engineers should be given particular responsibilities for various sections of the works i.e., drainage, buildings, roadworks, external works, etc.

A typical span of control for the site engineering staff on a large housing project is shown in Fig. 4.8.

Keeping track of progress

As siteworks are carried out it is essential that progress is accurately recorded. At least one member of the site engineering team should be present at the weekly site meeting, at which he will agree the future week's tasks and report operations completed to

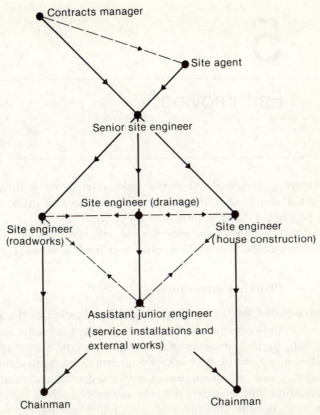

Fig. 4.8 Typical span of control and delegation for the site engineering
staff on a large housing project of say 300–500 dwellings

date. The progress of each section of the works should be entered
on the site programme, by colouring in each barline or critical
path operation.

One copy of each site layout plan for roads, drainage and buildings etc., should be reserved for progress purposes. As each section of the works is completed, the plan may be marked in various
colours e.g. on a drainage plan: blue for storm drains, red for foul
drains. On a housing site the house site layout plan can be shaded
to show when foundations are dug, concreted, bricked up, hard-
cored and oversite completed. Thus visual progress plans are
formed, giving accurate indications of those areas of work which
are ahead or behind programme.

5

Earthmoving

Chapter 4 reviewed the initial tasks carried out before earthmoving commences and the methods required to mark out soil storage heaps that need little control of finished level.

Chapter 5 reviews the procedure of bulk earthmoving on a contract of medium size where control over levels is necessary.

Planning earthmoving

The Bill of Quantities will give a good indication of the amount of excavation required on the project. Depending on the format of the bill, various amounts of excavation may be measured under different sections of the works. For instance, the substructures section may measure an amount of reduce dig and foundation trench excavation, but the external works and other sections may also do likewise. In order that the total measured excavation can be ascertained, a summary sheet must be drawn up, accumulating all relevant sections of the Bill of Quantities.

Whilst the measure will indicate the amount of excavation required, the layout of areas of excavation can only be determined by evaluation of the drawings. By calculating the difference between existing and final reduce levels, the depths and principle areas of excavation or fill can be discovered. It is useful when consulting the plans if areas of cut and fill are shaded using different colours to display the locations of each.

The area in which excavation will commence may depend on various factors e.g., the access to and from site (if the excavated material is to be removed), the programmed areas of completions, site conditions, relationship of cut and fill conditions etc. Following evaluation of all relative points, the planned method for excavation is determined together with the contracts manager and site agent.

Plant evaluation

Once the sequence of excavation areas has been agreed, the type and amount of plant required has to be considered. If the excavated material is to be removed from site, excavators loading into lorries will frequently be used, alternatively, if the spoil is to be retained and merely transferred from one area to another (i.e., from a cut to a fill area), motorized scrapers could be employed. The output rates of the equipment have to be found, which varies depending on the type of material requiring excavation, the distance to tip/fill area or spoil heap, the site conditions, time of year and climate.

Where lorries are needed for transferring material from site to tip, the mileage and route taken has considerable effect on the number of vehicles required. Enough lorries must be used to ensure that excavating machinery is kept at its optimum output capacity and yet no lorries should be kept standing between loads. The tip's location has therefore to be confirmed and an approximate round trip time calculated to determine the number of lorries necessary.

Throughout the project, amounts and types of earthmoving plant must be constantly reviewed, ensuring that maximum output is being achieved. As soon as the amount of equipment can be reduced the machine should be put off hire. From time to time more plant may be needed to boost output or equipment may need to be changed i.e., from a wheeled to a tracked machine, to cope with site conditions. It is easy to have plant on site ready for any eventuality but, although frequently seen in practice, this is extremely wasteful, costly and unnecessary.

An alternative to using the contractor's own or hired plant is to sub-contract the work to a specialist.

Subcontracting earthmoving

On sites where considerable earthmoving has to be carried out, the work may be placed in the hands of a specialist subcontractor. This generally occurs where the main contractor has either insufficient or no suitable machinery of his own capable of excavating the type of earthmoving required. Alternatively, the plant he owns may be in use elsewhere and the high cost of the equipment, together with the need for experienced operators, dictates that no additional plant is purchased and that a subcontractor be used.

Generally, a subcontractor will have tendered for the job by pricing the Bill of Quantity measured amounts and if employed on the works, may expect to excavate the quantities stated in one continuous operation. Unfortunately, this is seldom the case in practice as the Bill of Quantities will measure excavation throughout the job, much of which will not be carried out until the main construction has been completed. As an example, the areas of gardens around houses will not be finally graded until both house and main footpaths or roads have been constructed, or until such time as the reduced ground levels can be accurately established. Even where ground levels can be evaluated there is little point in reducing such areas to their final level when the ground will be subjected to constant traffic and material movement throughout the period of construction. It is therefore usual to leave a surcharge of earth in such places to allow for disturbance during the works, the surplus then being removed along with the subsoil grading of the areas. Obviously it is not always practical to expect the subcontractor to return equipment several months after the bulk earthmoving has been completed. The contractor may therefore choose to bring in his own machine for external works areas, as and when required. Should this occur, an accurate appraisal of the job at the intial stages should be made, with the subcontractor's order being placed to cover the exact quantities of bulk excavation required. Those areas of later excavation such as footpaths, gardens and parking bays etc., are then priced separately to cover the costs relative to the intended methods of excavation on site.

Setting out for earthmoving

Profiles

Excavation levels when earthmoving for roads, structures or external works are generally controlled by the use of profiles and a traveller.

The profile is a sight rail constructed of a timber cross board supported by one or two timber stakes (see Fig. 5.1). The size of the sight rail depends on the durability required and the distance over which sighting between profiles must take place. The sight rail will be set at a known level above the finished construction level or, if desired, above the final formation level. All sight rails should be set to enable easy sighting and are thus generally placed between 750 and 1800 mm above existing ground level, although occasionally this may not always be possible.

The traveller is used to establish the level of excavation, this

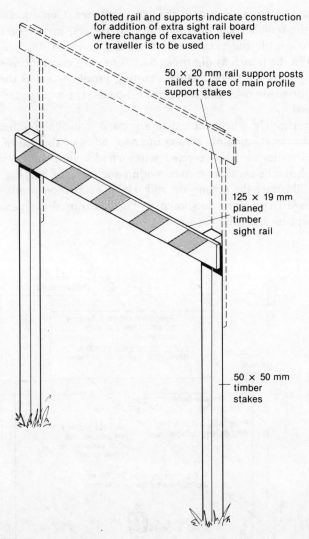

Dotted rail and supports indicate construction for addition of extra sight rail board where change of excavation level or traveller is to be used

50 × 20 mm rail support posts nailed to face of main profile support stakes

125 × 19 mm planed timber sight rail

50 × 50 mm timber stakes

Fig. 5.1 Twin stake profile construction for sights above 15 m in length

being achieved when the top of its cross board is found to coincide with the imaginary sightline viewed between the profile sight rails. (Chapters 6 and 7 discuss the methods of excavation level control for drainage and roadworks, although the principles of sight rail use remain similar throughout.)

During bulk excavation, which may be of considerable depth, the height of traveller must be kept to a manageable length to facilitate transportation over the excavated area. Consequently, to keep the traveller length to within four metres, it may be necessary to profile the excavation at various levels of dig, stepping down in the trench as dig progresses. On deep excavations where sloping sides are to be formed, 'cutting profiles' such as the type used for roadworks and described on pages 133 to 135, are often required.

Alternatively where the existing ground is well above final excavation level, digging may commence without the use of sight rails. Profiles are then erected when checks with the engineer's level show the excavation to be within one metre of final dig level.

To illustrate the setting up and use of excavation profiles, the proposed factory project shown on Fig. 5.2 is used as an example and explained as follows:

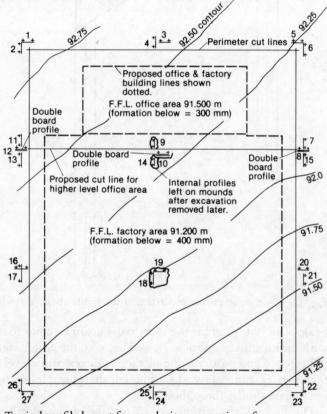

Fig. 5.2 Typical profile layout for use during excavation of a factory/office building

1 Establish depth of excavation required by comparison of existing ground levels and final excavation levels.
2 Determine if excavation may proceed to some depth prior to erection of profiles.
3 Consider whether cutting profiles should be erected to control sloping sides of excavation if drawings show this to be necessary.
4 Where sight rails are set to a distance above finish floor level, calculate the formation depth below finished floor of factory, and make this allowance on the traveller.
5 Decide on location and offset distance of profile stakes from factory.
6 Set out boundary lines of construction. Offset profiles to the required distance and drive in sight rail support stakes.
7 Using the engineer's level, mark the profile sight rail height on the stakes. The level will be relative to the traveller length and sighting height.
8 Nail on sight rails and decorate in distinctive colours, changing the colours to indicate where a new traveller length is required.
9 Construct a free standing traveller as shown in Fig. 5.3, ensuring that the total length is measured from the base of the support legs to the top of the traveller cross piece.
10 Mark out the required cut lines and proceed with the excavation.

Method of calculation for profile sight rail heights and traveller lengths for factory area in Fig. 5.2

Profile sight rail levels

It can be seen that the office area of the factory in Fig. 5.2 has a finished floor level (F.F.L.) of 91.500 m and it is from this that the sight rail height of the profile will be set. By inspection of the plan it can be seen that the highest existing ground level is at the corner where profiles 1 and 2 are situated, the level here being approximately 92.850 m. It is decided that the most convenient level for the sight rail to suit both sighting and traveller length will be to set this at 93.750 m. This allows 900 mm above ground at this point for sighting purposes with no other profile for the area (these being 1 to 12 inclusive) erected at a height greater than 1.70 m above existing ground level. The sight rails are therefore set at 2.25 m above the office F.F.L.

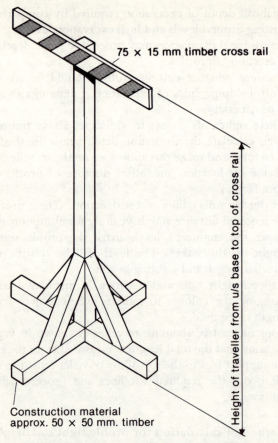

75 × 15 mm timber cross rail

Height of traveller from u/s base to top of cross rail

Construction material
approx. 50 × 50 mm. timber

Fig. 5.3 Typical detail of excavation traveller showing free standing
construction method

The sight rails used for the main factory area dig are to be set at a different height from the office area and relate to the factory F.F.L. The highest ground level is at the location of profiles 12 and 13, this being approximately 92.580 m. In an attempt to keep all sight rails at a constant height above the factory F.F.L., it is decided that a sight rail height of 93.200 m is to be used, although this will make profiles 12, 13, 22 and 23 slightly difficult for sighting height. However, as sights can be made to these from more conveniently levelled profiles, this level will be used. The sight rails for the factory are therefore set at 2.00 m above F.F.L. To aid identification, profiles in different areas should be painted in different colours. It can be seen that profiles 8, 10 and 12 are double board profiles, having sight rails for use in both areas (see Fig. 5.1).

Traveller length

The traveller used for finding the excavation level between profile sight lines must be constructed to equal the distance between sight rail and F.F.L., plus the distance below F.F.L., to allow for the formation of the floor.

The traveller length can therefore be calculated as follows:

Office area $= (93.750 - 91.500) + 0.300 = 2.25 + 0.3 = 2.55$ m
Factory area $= (93.200 - 91.200) + 0.400 = 2.00 + 0.4 = 2.40$ m

The traveller should be constructed in the same way as the example in Fig. 5.3.

Marking the excavation cut lines

Where it can be determined that large depths of excavation are required before the final reduce level is approached, the initial cut lines should be extended to make allowance for any battering or stepping of the excavation sides. The final cut lines are then defined when the formation summit is reached. The initial cut lines may be marked by range poles or long stakes. Further markers should be placed as the excavation proceeds to show the step or batter lines before the final excavation area (plus any additional working space required for scaffolding etc.) is reached.

Final cut lines may be set out using various methods e.g., taped distances from offset pegs, theodolite sightings along their edges, string lines or sight rail and guide lines etc. The method used will largely depend upon the type of construction and accuracy required. Frequently, actual cut lines are portrayed on the ground by spreading a thin layer of white lime below a string line. Excavation is then undertaken to correspond with the white edge lines, although these must frequently be checked by more accurate methods as the digging proceeds. On accurate work, where the excavated earth may for instance, form the edge shutter, final trimming by hand may be necessary. (Marking out trench excavation for drainage and structure works will be explained in Chapters 7 and 8.)

Controlling the excavation

The line and level of excavation will be controlled from the previously erected profiles, cut line markers or offset pegs. The type of plant required or subcontractor to be used will have been decided upon and the site engineer should be responsible for ensuring that plant operators are fully aware of the details for the excavation required. Before commencing excavation, the following criteria must therefore have been considered and explained to the plant operator:

1 the commencing point for excavation
2 areas to cut and fill
3 routes of any existing services running close by, across or over the site
4 working space required around construction area
5 spoil heap locations
6 method of marking cut lines
7 profiles to be used for various sections of the excavation
8 traveller lengths to be used for various sections of the excavation
9 sequence of excavation from the commencing point.

As excavation proceeds, spot levels should be taken using the engineer's level to check the excavation depths reached. During winter conditions, it may be prudent to leave an overburden of say 150 mm where it is unlikely that the area can be protected by a covering of hardcore or other material immediately after excavation. This overburden is then removed just prior to the laying of hardcore or finished material.

Consideration must be given to the type of plant required later for excavation of foundations etc. It can be seen by reference to Fig. 5.2 that the cut line for excavation between the factory and office areas protrudes 2 m into the area of the factory. This allowance is necessary as it is known that a tracked excavator will be used to dig the foundation between the factory and office. As the tracked machine must stand level when excavating, the ground is left at the **higher** level and reduced to the lower level by the tracked machine as the foundation dig proceeds.

Finally, as the excavation proceeds, any soft spots of ground discovered must be reported to the clerk of works, who will determine, together with the architect, if these are to be excavated to a solid base. The size and location of soft spots, where extra excavation is required, should be recorded and passed on to the site quantity surveyor, who will pursue payment for the work undertaken.

6

Roadworks

The setting out of roads presents the site engineer with an interesting challenge. In this chapter we shall investigate the techniques necessary for the locating of roads on site. The setting out of estate roads for large housing or industrial contracts is described, rather than the procedures to be undertaken for the layout of large motorway and civil engineering works. However, the latter is, in many cases, a scaled up version of the surveying techniques used for smaller roadwork.

The author considers that the best method of road setting out is to set only the critical centre line points i.e., tangent and junction positions etc., and offset the edges from these rather than peg the centre line at each chainage point. The intermediate chainage pegs should then be set at locations between each major centre line offset along each side of the road.

Practical experience has proved that, if the methods later explained are adopted, much unnecessary, time-consuming work can be avoided, without losing the accuracy of the final positions located in the field.

The correct locating and alignment of roads is of critical importance on any project. Errors may result in reduced plot sizes on housing schemes, encroachment on to adjacent boundaries, incorrect alignment of future road connections and numerous other equally onerous problems. The need for a high degree of accuracy and control is becoming increasingly important with the increase in high density housing and the need on most building land to use all the space available.

Time to construct and information available

It is often an advantage to construct the roadworks at the earliest possible stage (a) to provide good access to all areas of the works and (b) to enable the engineer to locate with less obstructions the setting out points required. Such practices will apply especially on housing schemes.

Obviously, before roadworks commence, other operations such as drainage and service installations (e.g., gas, water and electric mains etc.) must be carried out where they run within, or conflict with, the road kerb edge lines.

The engineer should ensure that adequate setting out information has been provided by the client's architect. The information should be clearly shown on all drawings and preferably include many, if not all, of the following items:

1 road centre line intersection points with co-ordinated or dimensioned reference information
2 road levels at centre and edge lines
3 co-ordinated and dimensioned radius centre points
4 tangent point locations with chainage dimensions
5 chainage distances from start to finish of the road section.

The amount of information provided will largely depend upon the extent of the roadworks to be constructed and the degree of accuracy stipulated. On projects where roadworks are fairly extensive, the architect normally issues information in the form of plans for definition of horizontal alignment, and longitudinal sections defining the vertical alignment.

Computer design is now becoming commonplace for the layout of road lines and although the architect may have issued his setting out drawings and longitudinal sections giving the major co-ordinated or detailed reference points, levels and radii, he could also have in his possession the computer print out containing even more information than that shown on the plans. Upon request, the architect will usually provide copies of the print out for use by the site engineer and this often saves calculation time.

Roadworks terminology

Before attempting to set out roads, the engineer must be conversant with the standard terminology used in road alignment and curve ranging. He will also need to learn the formulae used and the calculation procedure necessary, to obtain the surveying information required. A list of formulae commonly used during the setting out of roadworks is shown in Table 6.1 on page 116.

On most estate roads of average layout, the engineer will encounter:

horizontal curves	–	curves on plan
vertical curves	–	curves in section
road gradients	–	the slope ratios of the road
arc lengths	–	the length of a curve
chord lengths	–	the straight line distance between two points on a constant curve
tangent point	–	the point at which a straight line touches the curve from its apex
deflection angle	–	the angle subtended at the centre of the curve between lines to its tangent points
versed sine	–	the maximum height measured between the chord line and curve at the midpoint on the chord
radius	–	the length from the centre point to the edge of curve line
intersection points	–	the joint location of two straights of different bearings.

Many other terms may also be used to a lesser extent e.g., tangent distance, apex angles, tangential angles. Information to assist in these terms is given in the following diagrams.

Figure 6.1. Details the location of the main components of the curve.

Figure 6.2. Illustrates the following:

(a) Simple curve, i.e., a curve of constant radius
(b) Compound curve, i.e., two or more curves of different radii, the centre points of which are the same side of a common tangent
(c) Reverse curve, i.e., a curve formed by two adjoining arcs, often of different radii, the centres of which are on opposite sides of the common tangent.

Figure 6.3. Shows a horizontal curve together with its points and angles.

Figure 6.16. Denotes details of a vertical curve.

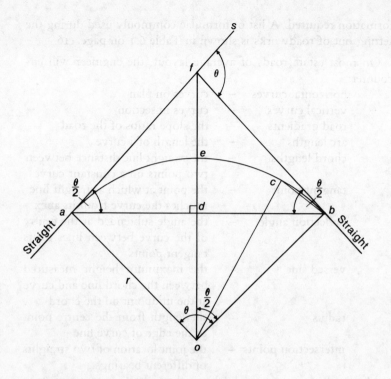

Line (*ab*) is termed the *long chord*
Line (*ed*) is termed the *versed sine* or *mid-ordinate*
Point *f* is termed the *intersection point* or *Apex*
Point *o* is termed the *centre of curvature for the radius*
Points *a* and *b* are the *tangent points*
θ = *deflection angle* = angle (*s* $\hat{f}$ *b*) = angle (*a* ô *b*)
$\theta/2$ = *total tangential angle* = ½ × *deflection angle*
Angle (*a* $\hat{f}$ *b*) = *intersection* or *apex angle* = 180° − θ
Line (*ef*) is termed the *external distance*
Line *r* is termed the *radius length*
Line (*ao*) = line (*ob*) = radius length
Line (*cb*) is a *short chord* on the curve *c* being at any point on the curve (*ab*)
Angle (*c* ô *b*) is termed the *deflection angle* for the *short chord*
Angle (*c b f*) is termed the *tangential angle* for the *short chord*
Angle (*c b f*) = ½ × angle (*c* ô *b*)
Point *d* is the *mid point* of the *chord (a d b)*
Point *e* is the *mid point* of the *curve (a e b)*
Lines (*af*) and (*bf*) are the *tangent lengths*

Fig. 6.1 Terms and relationships used during curve ranging

114

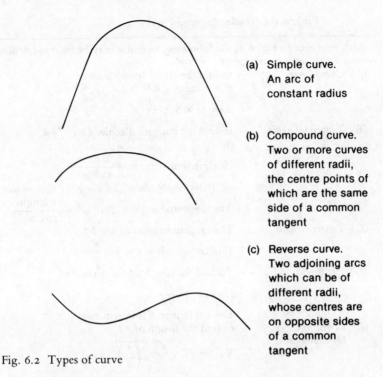

(a) Simple curve.
    An arc of
    constant radius

(b) Compound curve.
    Two or more curves
    of different radii,
    the centre points of
    which are the same
    side of a common
    tangent

(c) Reverse curve.
    Two adjoining arcs
    which can be of
    different radii,
    whose centres are
    on opposite sides
    of a common
    tangent

Fig. 6.2 Types of curve

Plotting the centre line reference pegs

Upon receipt of all available information, the engineer's first priority will be to calculate the angles and distances from the existing co-ordinate locations or grid points, to the centre line positions of the road. (The number of setting out points usually required on a large site and the methods of co-ordinate calculation is explained in Ch. 4.) The locating of these centre line pegs should be undertaken in the same manner as grid peg locations (also explained in Ch. 4).

Designers of the road scheme will have given co-ordinated values to reference points, which are usually located at the end of each straight and curved section of the road, thus denoting the centre line start and finish points of curves and straights known as the 'intersection points', or with a curve 'tangent points'. After the setting out information has been calculated and checked, the pegs marking these points can be positioned and protected, prior to offsetting to a known distance from the road edge lines.

Table 6.1 Formulae for roadworks

With reference to Fig. 6.1, the following formulae may be used in calculations when setting out curves.

(A) *Chord length* — to find the chord length $a\,d\,b$ or $(c\,b$ using $\angle\ c\,\hat{o}\,b)$

$$a\,b = 2r \times \mathrm{Sin}\ \frac{\theta}{2}$$

(B) *Tangential angle* — to find the tangential angles $f\,\hat{a}\,d$, $f\,\hat{b}\,d$ etc.

$$\text{'Sin' tangential angle} = \frac{\tfrac{1}{2}\ \text{chord}}{\text{radius}}$$

or 'Rankines Method' (*answer gives* $\angle$ *in minutes*)

$$\text{The tangential angle} = 1718.9 \times \frac{\text{arc length}}{\text{radius}}$$

(C) *Tangent length* — The tangent lengths $a\,f$ and $b\,f$

$$\text{Tangent length} = r \times \mathrm{Tan}\frac{\theta}{2}$$

(D) *Arc length* — To find the length of the curve $a\,e\,b$

$$L = 2\pi r \times \frac{\theta}{360}$$

$L = r\,\theta$ (where θ is in radians)

(E) *Versed sine (or mid-ordinate)* — to find the length of $e\,d$

$$\text{V.S.} = r - \sqrt{r^2 - \left(\frac{L}{2}\right)^2}$$

or

$$\text{V.S.} = r\left(1 - \cos\frac{\theta}{2}\right)$$

Where L = chord length $a\,d\,b$
r = radius $a\,o$

(F) *Deflection angle* θ — To find the angle $a\,\hat{o}\,b$

$$\theta = \frac{L}{r} \qquad \text{where } L = \text{arc length } a\,e\,b$$
(θ given in radians)

(G) *External distance* — To find the length $f\,e$

$$\text{External distance} = r\left(\sec\frac{\theta}{2} - 1\right)$$

On occasion, the client's representative may check all or a random number of the engineer's calculations for road setting out points as well as their physical location when fixed on site. This in itself provides a double check for the engineer's work. However, the engineer, as a precaution, should take measurements of the lengths between pegs to prove the accuracy of the work when compared with the chainage length for straights shown on the plans, and the calculated chord lengths for the curves.

When all checking of the road centre line points is complete and proved correct, the engineer should proceed with the offsetting of the centre points to each side of the road. These offset pegs then provide the basis for marking of the edge lines used in road excavation and re-location of the centre line point when required. Offsets should be placed at a standard distance from face of the proposed kerb line e.g., 1 m, or alternatively clear of any adjacent footpaths, dig or drainage etc., which may be undertaken in conjunction with, or during the course of, road construction.

Setting out the road offset pegs

As explained, the offset pegs should be located at a known distance from face of kerb and are therefore positioned at a length equal to half the road width plus the required offset to each side of the centre line. This distance will vary according to the amount of excavation required and the extent to which the road edges need to batter back i.e., the deeper the road, the farther away the offset must be placed. The engineer places his theodolite on the centre line peg and turns the required angle after sighting a reference peg of known location. In the case of a straight this will be 90° from the straight's centre line, for a curve often ½ × (180° − the deflection angle) or by sighting the previously located radius point centre peg. Figure 6.3 shows the procedure in detail.

As shown, a theodolite can be set on points A or B and offsets placed by one of three ways.

1 By sighting along the centre line of each straight, turning 90° and measuring half the road width plus the offset distance each side to locate C, D, E and F.

2 By sighting B from A and turning $\dfrac{180° - \theta}{2}$ and measuring as before.

3 By sighting back to the radius peg O from A or B and measuring as before.

When the offsets to the main centre lines are fixed, the engineer must locate intermediate pegs on the required points between main offsets to curves and straights. Usually and preferably these points are fixed to correspond with chainage positions and given road levels. In the case of even grade straight roads, pegs can be at every 20 m, but where vertical curves are detailed, closer points will be necessary e.g. at every 5 or 10 m depending on the curve

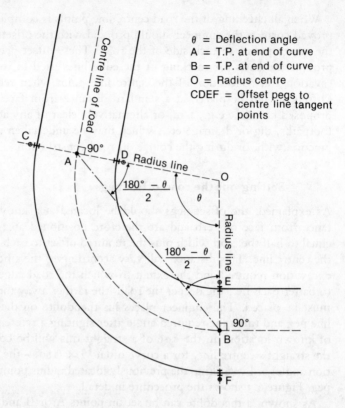

Fig. 6.3 Methods of offsetting centre line pegs from straights and curves

gradients. Similarly, horizontal curves on a large radius may have points every 10 m with smaller radii having pegs at 5 m intervals. It is extremely important to remember that a curve with chainage points shown at 5 or 10 m on the centre line will have a reduced arc length when offset on the inner curve and a greater arc length when located on the outer curve. Deflection angles and tangential angles of each arc will however remain constant throughout the segment in question.

Obviously when taping between curve offsets, the engineer needs to calculate and use a chord length distance (i.e., the straight line distance between points) as shown in Fig. 6.1, line *é b*. Chord lengths for inner and outer radii are calculated using the formula listed in Table 6.1.

Calculation and record sheet for road curves

Contract.................   Date.................

Point no	℄ Ch.M	℄ Arc length	Deflection angle	Tangential angle	Chord length		Chord length adj: for scale factor		Versed sine		Versed sine adj. for S.F.		Information on curve
					Inner	Outer	Inner	Outer	Inner	Outer	Inner	Outer	
													Between points on road ....
													Centre line radius ....
													Inner curve radius ....
													Outer curve radius..........
													Scale factor used ....
													Notes and remarks
Totals													

Fig. 6.4 Calculation and record sheet for road curves

Calculating for curve setting out

The engineer will save time if a programmable pocket calculator is used and the calculation work is recorded for reference on standard printed calculation sheets. A pro-forma for such a sheet is detailed in Fig. 6.4.

Distances for centre lines, road kerb lines, offset lines and angles etc., can be calculated in a running program devised to give the information required using the details available. Examples of programs formulated for this purpose to be used on a Texas Instruments TI57 programmable calculator are given later in this chapter, but once the operator has become conversant with the methods of programming, similar programs can easily be compiled for use with other programmable calculators.

Before starting any calculations, checks should be made to ensure that adequate information is available i.e., such details as tangent points, radius lengths and centre point locations, chainage lengths and road levels. Ideally details should also include data on arc lengths, deflection angles and vertical curves together with co-ordinate references. The amount of information given will, of course, vary from job to job and one site may involve more calculation time than another. Occasionally, the architect may only provide co-ordinate centre points and radius length details for curves. In such cases extra calculation work will be required to establish deflection angles and arc lengths by intersection of tangents and bearings. The architect should advise if a local scale factor is to be used to correct the National Grid, and if so procedures for its calculation should be followed as described in Chapter 4. Where co-ordinates are not used the architect will have to define the road centre line by an alternative method.

To establish the procedures involved in curve setting out, a series of worked examples on the following pages will assist the reader in his understanding of the methods best adopted.

Setting the offset pegs to each side of the curve

As stated on page 117 the offset pegs adjacent to the main centre of road points will have been located first and intermediate pegs have to be placed to correspond with the chainaged distances. This procedure is preferably carried out by using one theodolite to swing angles and a tape to measure distances between pegs. With reference to Fig. 6.5 and its corresponding calculations sheet Figure 6.6 evaluation of these procedures will commence.

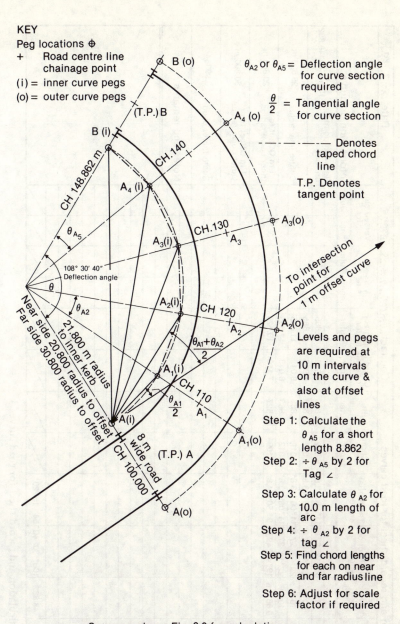

KEY

Peg locations ⊕
+   Road centre line chainage point
(i) = inner curve pegs
(o) = outer curve pegs

θ_{A2} or θ_{A5} = Deflection angle for curve section required

$\frac{\theta}{2}$ = Tangential angle for curve section

—·——·—— Denotes taped chord line

T.P. Denotes tangent point

B (o)

(T.P.) B

A_4 (o)

B (i)

CH.140

CH 148.862 m

A_4 (i)

θ_{A5}

$A_3(i)$

CH.130

A_3

$A_3(o)$

108° 30′ 40″ Deflection angle

To intersection point for 1 m offset curve

θ

θ_{A2}

$A_2(i)$

CH 120

A_2

$A_2(o)$

Near side 20.800 Far side 30.800 radius to offset
21.800 m radius to inner kerb
radius to offset

$\frac{\theta_{A1}+\theta_{A2}}{2}$

$A_1(i)$

CH 110

A_1

$A_1(o)$

$\frac{\theta_{A1}}{2}$

A(i)

8 m wide road

(T.P.) A

CH 100.000

A(o)

Levels and pegs are required at 10 m intervals on the curve & also at offset lines

Step 1: Calculate the θ_{A5} for a short length 8.862
Step 2: ÷ θ_{A5} by 2 for Tag ∠
Step 3: Calculate θ_{A2} for 10.0 m length of arc
Step 4: ÷ θ_{A2} by 2 for tag ∠
Step 5: Find chord lengths for each on near and far radius line
Step 6: Adjust for scale factor if required

See example on Fig. 6.6 for calculation

Fig. 6.5 Setting out of curve from 1 m offsets using one theodolite and taped distances. Given θ (deflection angle), radius length and chainage for at least one end, curve can be set out by taped lengths and sighting of tangential angles

Calculation and record sheet for road curves   Contract HARTOP HOUSING ESTATE   Date 15.7.80

Point no	Ch.M	Arc length	Deflection angle	Tangential angle	Chord length Inner	Chord length Outer	Chord length adj. for scale factor Inner	Chord length adj. for scale factor Outer	Versed sine Inner	Versed sine Outer	Versed sine adj. for S.F. Inner	Versed sine adj. for S.F. Outer	Information on curve
A	100 m												Between points on road CH 100–CH 148.862
A1	110 m	10 m	22°12'27.6"	11°06'13.8"	8.012	11.863	8.013	11.865					
A2	120 m	10 m	22°12'27.6"	11°06'13.8"	8.012	11.863	8.013	11.865					Centre line radius ..... 25.800 m
A3	130 m	10 m	22°12'27.6"	11°06'13.8"	8.012	11.863	8.013	11.865					1 m Offset to Inner curve radius ..... 20.800 m
A4	140 m	10 m	22°12'27.6"	11°06'13.8"	8.012	11.863	8.013	11.865					
B	148.862	8.862 m	19°40'49.6"	9°50'24.8"	7.109	10.527	7.110	10.529					1 m Offset to Outer curve radius ..... 30.800 m
													Scale factor used 0.99984
													Notes and remarks
Totals		48.862	108° 30' 40"	54° 15' 20"									

Fig. 6.6 Calculation and record sheet for road curves

The required pegs for this curve are to be at 10 m intervals (centre line arc distance) and should commence at chainage 100 m and complete at chainage 148.862 m. Corresponding pegs need to be placed at a known offset distance to the edge lines (say 1 m) thus involving calculation of the chord lengths and deflection angles. It is evident that four number 10 m chainages are required plus a completing arc of 8.862 m along the centre line, with corresponding pegs located at the offset lines. Remembering that the deflection angles will be indentical at centre and edge lines commence calculation.

Step 1 Find the deflection angle for an 8.862 m arc by use of a programmable calculator and enter value on pro-forma sheet (see Fig. 6.6 and Table 6.1 for formulae).

As tangential angles will be turned by the theodolite:

Step 2 Is to divide the deflection angle by 2, as a tangential angle = ½ × deflection angle.

Step 3 Undertake the same procedure as step 1, but for a 10.0 m arc.

Step 4 Undertake the same procedure as step 2, but for a 10.0 m arc.

As chord lengths are the measured lengths required to be taped:

Step 5 Involves the calculation of chord lengths for offsets at both sides of the road (see Table 6.1 for formula and Fig. 6.7 for calculator program).

Record the calculated lengths on the pro-forma. Finally adjust the lengths by a scale factor if required prior to setting out on site.

When on site, locate the theodolite on point A (i) or A (o), sight point B (i) or B (o), and set the theodolite scale to the total tangential angle. Depending on the type of curve i.e., right hand or left hand, angles may have to be deducted from 360° due to the location of the theodolite and direction of swing.

In the case of the curve shown on Fig. 6.5, the angles should be deducted from 360°, therefore when setting on A (i) and sighting B (i) the theodolite must be pre-set on the angle resulting from the following calculation:

$$\text{Total tangential angle} = 360° - \frac{(\text{deflection angle})}{2}$$

$$\therefore = 360° - \frac{(180° \ 30' \ 40'')}{2}$$

$$\therefore \quad = 360° - 54° \ 15' \ 20''$$

$$\therefore \quad = 305° \ 44' \ 40''$$

Likewise those angles for the intermediate points will each have to be deducted from 360°.

With reference to Fig. 6.6 (which uses the calculation for the road curve sheet of Fig. 6.4,) it can be seen that the total tangential angle is made up of the sum of all other tangential angles. Therefore, when plotting the curve on site, each angle is added in turn and here deducted from 360°. For example, after presetting on point B at 305° 44′ 40″ the angle to point:

$A1 = 360° - 11° \ 06' \ 13.8'' = 348° \ 53' \ 46.2''$
$A2 = 360° - 22° \ 12' \ 27.6'' = 337° \ 47' \ 32.4''$
$A3 = 360° - 33° \ 18' \ 41.4'' = 326° \ 41' \ 18.6''$
$A4 = 360° - 44° \ 24' \ 55.2'' = 315° \ 35' \ 04.8''$
$\ B = 360° - 54° \ 15' \ 20'' \quad = 305° \ 44' \ 40''$

As previously stated, the same angles will be used on both sides of the curve where it has a concentric radius.

Chord lengths will have been calculated for both inner and outer offset radii, as obviously these lengths differ as the curve increases or decreases in arc size. The calculation for the chord length is by use of the formula given in Table 6.1.

The chord lengths for the curve as detailed on Fig. 6.5 will therefore be for the centre line arc lengths of 10 m and 8.862 m, when transferred to the inner offset radius of 20.8 m, and outer offset radius of 30.8 m.

An example of the calculation for the 10-metre arc chord length on the inner radius would therefore be as follows:

$$\text{i.e. chord length} \quad = 2R \times \mathrm{Sin}\dfrac{\theta}{2}$$
$$\text{chord length} \quad = 2 \times 20.8 \ \mathrm{m} \times \mathrm{Sin} \ 11° \ 06' \ 13.8''$$
$$\therefore \quad = 41.6 \ \mathrm{m} \times 0.192 \ 587 \ 6$$
$$\therefore \quad = 8.012 \ \mathrm{m} \ \text{(to the nearest mm)}$$

Likewise the same 10-metre arc on the 30.8 m outer radius would have a chord length of:

$$\text{chord length} \quad = 2 \times 30.8 \ \mathrm{m} \times \mathrm{Sin} \ 11° \ 06' \ 13.8''$$
$$\therefore \quad = 61.6 \ \mathrm{m} \times 0.192 \ 587 \ 6$$
$$\therefore \quad = 11.863 \ \mathrm{m} \ \text{(to the nearest mm)}$$

The same procedure is adopted for the 8.862-metre arc and the results of this arc are shown entered in the record sheet on Fig. 6.6.

Program record sheet for Texas TI 57 calculator

Program title:	To calculate the chord length of an arc given R (radius) and θ (deflection angle)

Program drawn up by:	R.W. MURPHY	Date:	JUNE 1979

Formulae used: CHORD LENGTH $= 2R \times \sin \dfrac{\theta}{2}$

Program description and notes

       The program includes for converting degrees–minutes–seconds into decimal degrees, therefore the angle can be entered direct as read from the drawings or calculation sheets.

Step number	Press	Key code	Calculator display	Step number	Press	Key code	Calculator display
1	LRN	—	00 - 00				
2	RCL 1	33 -1	01 -00				
3	2nd D M S	26	02 -00				
4	÷	45	03 -00				
5	2	02	04 -00				
6	=	85	05 -00				
7	2nd SIN	28	06 -00				
8	×	55	07 -00				
9	2	02	08 - 00				
10	×	55	09 - 00				
11	RCL 2	33 -2	10 - 00				
12	=	85	11 - 00				
13	R/S	81	12 - 00				
14	LRN	—	0				
15	RST	71	0				

Information required in memories fed into:
STO 0 –
STO 1 – Enter θ (in D.M.S) deflection angle
STO 2 – Enter R (radius) value in metres
STO 3 –
STO 4 –
STO 5 –
STO 6 –
STO 7 –

Fig. 6.7 Program record sheet for calculation of chord lengths

When undertaking a large number of simultaneous calculations for chord lengths, time savings can be achieved by the use of a progammable calculator fed with a suitable programme. An example of a TI57 programme is given in Fig. 6.9 – this being drawn up to include the entry of the deflection angle direct.

If plans have been prepared from the National Grid system, adjustment of the chord lengths by a local scale factor to convert to ground distance from the National Grid may be required and stipulated by the client. The engineer will have to establish the value of the local scale factor by reference to tables or by calculation and adjust the chord lengths accordingly.

In Fig. 6.6, the scale factor used has been stated as being 0.999 84. The calculation to convert projection distances (i.e., the distances calculated from two ordnance survey stations) to the actual ground distance measured is:

$$\text{ground distance} = \frac{\text{projection distance}}{\text{local scale factor}}$$

Therefore, for an 11.863-metre chord on the outer radius of the curve calculated in Fig. 6.6, the ground distance $= \dfrac{11.863}{0.999\ 84}$

$\therefore$ Ground distance $= 11.865$ m (to nearest mm)

The engineer now has adequate information to locate on site his curve line offset pegs for use in profiling, excavation and final positioning of the kerb lines.

If desired, and where taping of distances would be difficult, two theodolites can be used and angles calculated from point B when sighting point A (refer to points B and A in Fig. 6.5).

To set out a curve when only co-ordinated radius points, radius lengths and straight bearings are given

Where the architect has only indicated the bearings of straights, co-ordinated intersections, radii and given radius lengths, the engineer will have to calculate arc lengths and deflection angles before continuing with calculations to set out the road curves as explained in the previous section.

Figure 6.8 shows an example where the road designer has given information on the drawings to denote the radius co-ordinates at point A and B, the bearings of the straights leading to points C and D, and the radius lengths for the curves originating at points

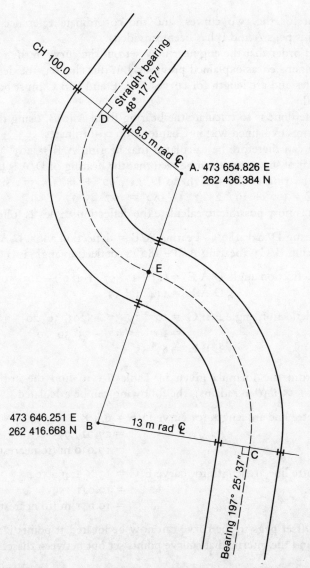

Fig. 6.8 Example of road curves to be set out using co-ordinated radius pegs

A and B. All other information required for the curves D to E and C to E will have to be calculated as follows.

Calculations for curve information on Fig. 6.8
As indicated on Fig. 6.8, only the bearings of the two straights the

radii for the two curves, and the co-ordinate reference of the radius pegs A and B has been provided.

In order that the engineer may set out the curve in the required manner i.e., as explained previously in this chapter, the deflection angles and arc length for curves D Â E and E B̂ C must be calculated.

Step one is to calculate the bearing from A to B, using the procedures explained within Chapter 4 on co-ordinates.

It can therefore be calculated that bearing A/B = 203° 30′ 20″. From observation it can be seen that the bearing of D/A is 90° plus the bearing of the straight to D, i.e., 90° + 48° 17′ 57″, similarly B/C = (90° + 197° 25′ 37″) − 180° = 107° 25′ 37″.

It is now possible to calculate the deflection angles as follows:

(bearing D/A + 180°) − bearing A/B = deflection angle D Â E
bearing B/C − (bearing A/B − 180°) = deflection angle E B̂ C

$$\therefore \text{ deflection angle D Â E} = (138° \, 17′ \, 57″ + 180°) − 203° \, 30′ \, 20″$$
$$\therefore \text{D Â E} = 114° \, 47′ \, 37″$$

$$\therefore \text{ deflection angle E B̂ C} = 107° \, 25′ \, 37″ − (203° \, 30′ \, 20″ − 180°)$$
$$= 107° \, 25′ \, 37″ − 23° \, 30′ \, 20″$$
$$\therefore \text{E B̂ C} = 83° \, 55′ \, 17″$$

From the formula given in Table 6.1 to find the arc length L = $R \times \theta$ (θ in radians), the following can be calculated:

Centre line arc length for curve D/E = 8.5 × 2.003 526 5
                                    = 17.029 975
                                    = 17.030 m (to nearest mm)

Centre line arc length for curve E/C = 13 × 1.464 704 5
                                    = 19.041 159
                                    = 19.041 m (to nearest mm)

Offset pegs to the curve can now be located at points D, E and C, and the intermediate curve points set out between these.

Summary of examples given

From the information given on the previous pages, the engineer will have reviewed examples of setting out roads from the main centre line points to offset pegs at required locations on straights and curves.

Program record sheet for Texas TI 57 calculator

Program title: TO FIND THE DEFLECTION ANGLE OF A CURVE – ARC LENGTH OF A CURVE – RADIUS OF A CURVE – CHORD LENGTH OF A CURVE.

Program drawn up by: R.W. MURPHY   Date: JUNE 1980

Formulae used: $\theta = \frac{L}{R}$, $L = \theta \times R$, $R = \frac{L}{\theta}$, $CL = 2R \times \sin\left(\frac{\theta}{2}\right)$ where θ = Deflection Angle, L = Length, R = Radius

Program description and notes

Given the arc length and radius θ can be found using steps 1–13 (label 0)
Given the radius and θ, enter θ into STO 3, enter R into STO 2 (label 1) use GTO 1 if desired
Given the arc length and θ, enter θ into STO 3 and L into STO 5 (label 2) use GTO 2 if desired
Given R and θ, enter R into STO 6 and θ into STO 4 (label 3) GTO 3 to find CL if desired

If GTO is used each labelled section can be called up or alternatively, program can be run throughout. (when G.T.O instruction is used R/s must be pressed to run the program from label №)

Step number	Press	Key code	Calculator display	Step number	Press	Key code	Calculator display
1	LRN	–	00 – 00	26	÷	45	25 – 00
2	RCL 1	33 – 1	01 – 00	27	RCL 3	33 – 3	26 – 00
3	÷	45	02 – 00	28	=	85	27 – 00
4	RCL 2	33 – 2	03 – 00	29	STO 6	32 – 6	28 – 00
5	=	85	04 – 00	30	R/s	81	29 – 00
6	STO 3	32 – 3	05 – 00	31	2nd LBL 3	86 – 3	30 – 00
7	×	55	06 – 00	32	RCL 4	33 – 4	31 – 00
8	1	01	07 – 00	33	÷	45	32 – 00
9	8	08	08 – 00	34	2	02	33 – 00
10	0	00	09 – 00	35	=	85	34 – 00
11	÷	45	10 – 00	36	2nd SIN	28	35 – 00
12	2nd π	30	11 – 00	37	×	55	36 – 00
13	=	85	12 – 00	38	2	02	37 – 00
14	STO 4	32 – 4	13 – 00	39	×	55	38 – 00
15	Inv 2nd D.M.S	– 26	14 – 00	40	RCL 6	33 – 6	39 – 00
16	R/s	81	15 – 00	41	=	85	40 – 00
17	2nd LBL 1	86 – 1	16 – 00	42	STO 0	32 – 0	41 – 00
18	RCL 3	33 – 3	17 – 00	43	R/s	81	42 – 00
19	×	55	18 – 00	44	LRN	–	0
20	RCL 2	33 – 2	19 – 00	45	RST	71	0
21	=	85	20 – 00				
22	STO 5	32 – 5	21 – 00				
23	R/s	81	22 – 00				
24	2nd LBL 2	86 – 2	23 – 00				
25	RCL 5	33 – 5	24 – 00				

Information required in memories fed into:
STO 0 – Enter or recall chord length
STO 1 – Enter arc length L
STO 2 – Enter radius length
STO 3 – Enter θ (in radians)
STO 4 – Enter θ (in decimal degrees) for chord length calculation
STO 5 – Enter arc length L for radius calculation
STO 6 – Enter radius length for chord length calculation
STO 7 –

Fig. 6.9 Program record sheet for calculation of deflection angle, arc length, radius and chord length of a curve.

It has been stressed that the use of a programmable pocket calculator can give fast and accurate solutions to the calculations required. The examples given are a guide to the methods of programming for a particular calculator. Once the engineer has become proficient in the use of his particular calculator, he will find that by adopting more complex techniques he can reduce further the processing time for calculations by means of more advanced programs i.e., the use of 'labels' for various sections of the program. The operator, by adopting a 'labelling' method, may choose to run a particular section of the program he has entered into his machine or alternatively, obtain all the relevant results for a calculation by feeding in information to the various calculator memories and 'running the program'.

An example of such a program for use with the Texas TI57 calculator is given on Fig. 6.9. This program enables the operator to

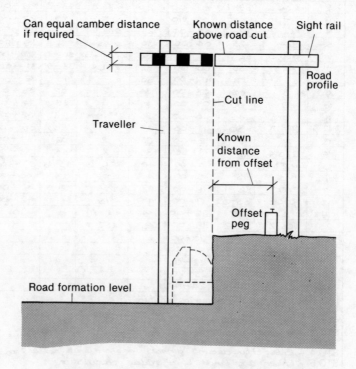

Road profile set to correspond with the required cut line in order that losses can be reduced or kept to a minimum

Fig. 6.10   Example of road profiles set for cut lines and levels

obtain easily such information as the deflection angle, arc length, radius length and chord lengths of a curve, once the program instructions have been placed in the machine.

Road level profiles

As the setting out of the road proceeds with establishment of the offset chainage location pegs, the engineer should erect road level profiles alongside these and any intermediate points at which levels are required. The profile sight rails will be set at a constant distance above road finish or reduce levels and be the guidelines for excavation of the road.

Sight rails are timber boards of say 75 × 15 mm section, set horizontally on to long stakes to be used in conjunction with a 'traveller'. The levels of excavation can be obtained by sighting

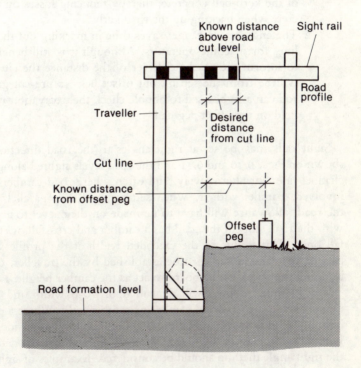

Road profile set to correspond with the required cut line formed by base of traveller set to provide excavation cut line

Fig. 6.11   Example of road profiles set for cut lines and levels

between profiles until the traveller crossboard coincides with the sight line between the sight rails (see Figs. 6.12 and 6.13).

When positioning the profile stakes at chainage locations, care should be taken to avoid disturbing previously located chainage point offset pegs. It is good practice and more accurate to position the sight rails not only to level, but also to have their roadside edge finished to coincide with the cut line of the excavation or alternatively at a set dimension from it, an allowance then being made on the traveller to denote the cut line.

Examples of both methods are given in Figs. 6.10 and 6.11.

Using the sight rails to denote cut lines may be slightly more time consuming to establish, but has advantages in two respects.

1 If the excavation can be cut to the true back of kerb base line, then it should only be necessary to shutter one face of the kerb bed concrete, thereby reducing losses on concrete when backing up the road kerbs.

2 The engineer or ganger saves time in marking out the cut lines for machine operators. Although it is still beneficial to inform the machine driver of the distance the cut line will be from the setting out offset pegs, a pre-cut gauge rod can also be used to double check the excavation when held on the offset peg nail.

Sight rails may be set at right angles to the road direction as shown on Figs. 6.10 and 6.11 and the road levels sighted along the channel lines, or alternatively, and often where road cambers are involved, parallel with it. When setting sight rails parallel with the road, allowance will have to be made on their level to accord with the road slope if the road has a camber and crossfall or crossfall only. This is due to the extended width at the profile stake offsets, the camber can then be established by the traveller, often by using the width of the board to act as the camber height.

The method adopted will depend upon road width and type i.e., crossfall, crossfall and camber or camber only. Where a camber only road is specified and channel lines are level each side, the parallel method is usually used. Where a crossfall road is detailed the right angle method should be considered. Examples of sighting and using each method are shown on the sketches on Figs. 6.12 and 6.13.

To assist in the identification of peg and profile points it is advantageous to mark each with a thick felt pen to show the

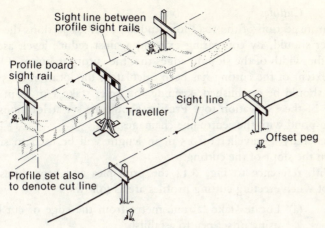

Fig. 6.12 Road profiles set at right angles to the road lines

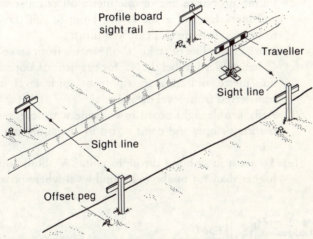

Fig. 6.13 Road profiles set parallel to the road lines

chainage distance and height of profile above road finish or reduced level.

Profiles for cuttings and embankments

When undertaking the construction of roads with deep excavation or excessive fill, the engineer will have to adopt a system of sloping (or batter) profiles set back from the edge of the excavation lines and used as sights up or down to establish the banking lines at the road verges.

Cuttings

From inspection of the road longitudinal or cross sections the engineer should, by comparing existing against reduce level, ascertain the width of the sloping edges that the cutting requires. Once the extent of the cutting has been calculated, the location for profiles should be established, keeping the inner stake 1 m from the edge of the excavation top. Profiles should be set at locations to correspond with the centre line chainages.

Generally a traveller of 0.5 m in length will be used to sight down the slope of the cutting.

With reference to Fig. 6.14 the principles and procedures to adopt when erecting cutting profiles are as follows:

(a) Locate stake '2' one metre from the edge of cut line using offset peg to establish.

(b) Locate stake '1' one metre from stake '2'.

(c) Plumb up centre line of one metre offset on stake '2' and mark level of grade at this point to suit traveller to be used (i.e., in this case 500 mm).

(d) Mark centre line of stake '1' one metre from stake '2'.

(e) Calculate vertical height 'C' for gradient of slope over one metre and measure up from spirit level mark transferred from level on stake '2'.

(f) Affix profile sight board to coincide with level points marking slope, and paint board in distinguishing colours.

(g) To assist in sighting through, point 'A' should not be higher than 1.5 m above ground level where possible.

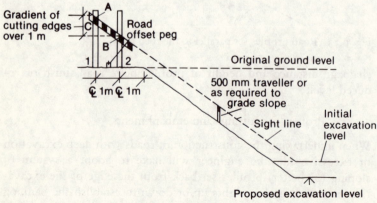

Fig. 6.14 Example of profiles for cuttings

When excavating a cutting it is generally felt beneficial to excavate the road to an initial level of say 150 to 300 mm above the final reduce level. From this stage the engineer will have to re-set out the road in the normal manner at the cut level to give more accurate control over the final excavation and cut lines.

Embankment and deep fill conditions

These are controlled from the setting out aspect in much the same way as cuttings, the obvious difference being that sight lines are produced upwards as opposed to downwards.

Figure 6.15 shows a typical embankment cross section. The profile should be constructed in the same manner as that described for cuttings, with an effort to make the lower edge of the sight board a minimum of 750 mm above existing ground to facilitate ease of sighting. Obviously, in this instance a larger traveller will be used and the road filled in consolidated layers to a height giving an overburden of approximately 150 mm. Once this level has been reached, the road will be re-set out in the traditional manner on top of the fill, then excavated to the reduce levels and accurate cut lines as required.

In cases where the fill is excessively deep, say over 5 m intermediate profiles will have to be constructed on the sloping embankment sides, ensuring accurate levels are maintained.

Vertical road alignment

As far as the site engineer is concerned, the vertical alignment of

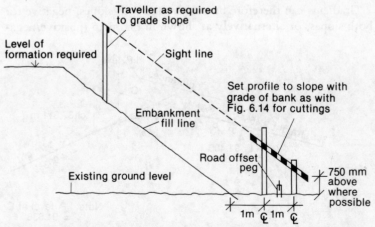

Fig. 6.15 Example of profiles for embankments and fill conditions

the road is fairly straightforward in so far as most of the information is usually given on the longitudinal sections of the road, and in the case of straight grades the engineer can easily interpolate a level required at any point.

The situation does become more complex when a section of the road is formed by a vertical curve. In most cases the designer will give the required road levels on the centre line and these may be placed at 10 m or 5 m intervals, depending on the grade of the curve. Alternatively, the drawings may only show the curve's start and finish levels, road grades and intersection point.

Whichever method has been presented, the engineer may wish to know the level of the road at points other than those shown. It will therefore be necessary for him to learn and practise methods of calculation for vertical curves of which there are various examples.

Vertical curves

A vertical curve is used to join two intersecting straight lines in a vertical plane and thus requires calculation of levels at various points throughout its length for establishment on site.

Figure 6.16 shows a vertical curve which is connecting two straights of known gradient (gradient being the ratio of slope usually stated as a percentage). During calculations the algebraic difference in gradients will be used.

> *Positive gradients rise from left to right*
> *Negative gradients fall from left to right*

Gradients can therefore be positive for both slopes, negative for both slopes, or alternatively as shown in Fig. 6.16 (positive/nega-

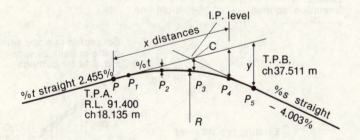

Note: I.P. level at C
= 91.638

Fig. 6.16 Typical vertical curve

tive) or where a dip occurs (negative/positive). The algebraic difference between gradients will also have a varying positive/negative result, depending upon the curve being in rise or sag and also between the difference of the ratios of slope for each straight. This negative/positive value can be ascertained visually from the long section relevant to the road curve to be calculated.

Calculation for the arc is based on the 'parabolic curve' and in the most common type of curve tangent lengths are equal (although occasionally unequal tangent length curves are designed).

It should be stressed that certain assumptions are usually made to simplify the calculations of the curve, providing both gradients are less than 4 per cent the most important of these being:

   1 the chord length = the arc length = the sum of the tangent lengths
   2 the tangent length = the horizontal length
   3 dimension y is regarded as being truly vertical between the entry tangent line and the curve.

Calculation example

In Fig. 6.16 it can be seen that the radius length R is not given. *R* can be calculated by using the formula:

$$R = \frac{\text{distance between tangent points } (L) \times 100}{\text{algebraic difference of the gradients } (A)}$$

$$\therefore R = \frac{L \times 100}{A}$$

With reference to Fig. 6.16, to obtain the level values for the curve's arc, distances are calculated to give the dimensions below line A/C at any distance along its length.

The value of the drop (or in the case of some sag curves, rise) is given by the formula:

$$y = \frac{(A) \, x^2}{200 \, L}$$

Where y = drop distance
       x = the distance along line A/C

Levels are calculated for points corresponding to the chainage positions to be plotted on site (e.g., 5 m intervals)

To establish the value of the curve's levels relative to the T.P. level, the reduced level (R.L.) is first calculated at the chainage point on the line A/C at distance x along its length and then y deducted (or sometimes added in sag curves) to give the level on the curve.

Level of curve at P, P_1, etc = R.L. at $x - y$ value, for each point

Where R.L. = level at T.P. 'A' + $\left(\dfrac{x \times \%t}{100} \right)$

and $y = \dfrac{(A)\, x^2}{200\, L}$

The levels are calculated from one end of curve e.g. T.P. 'A' at level 91.400. However, if only the I.P. level at point C is given, T.P. 'A' can be calculated by the equation:

$$\text{T.P. 'A'} = \text{I.P. level} - \left[\frac{(A/C) \times \%t}{100} \right]$$

$$\therefore \text{T.P. 'A'} = 91.638 - \left[\frac{9.688 \times 2.455}{100} \right]$$

$$\therefore \text{T.P. 'A'} = 91.638 - 0.238$$

$$\therefore \text{T.P. 'A'} = 91.400 \text{ m}$$

During calculations it is helpful to use a pro-forma to record the values of computation and reduce level for easy reference during setting out and for recording in the calculation file. Such a pro-forma compiled with the results for the vertical curve in Fig. 6.16 is shown in Fig. 6.17.

Three centre compound curves

As such curves are not very frequently specified in general estate roads, lengthy investigation of their complex compilation is not felt desirable within the scope of this book. However, the basis of calculation for this type of curve together with the setting out details can be researched in more detailed books specifically relating to roadworks, and tables for quick calculation are available (see endnote 1).

Spiral transition curves

Calculation for these curves has been made easier by the publication of the *Highway transition curve tables* printed on behalf of the County Surveyors Society. As this type of curve is only oc-

Information	Chainage	x for line A–C	RL at x T.P.A. + $\frac{(x \times \%t)}{100}$	$y = \frac{(A) x^2}{200L}$	$R = \frac{L \times 100}{\text{'A'}}$	Reduce level point 'P'
R = 300.031 m	18·135	0	91·400	0	300·031	91·400 p
%t = +2.455%	20·000	1·865	91·446	−0·006	"	91·440 p1
	25·000	6·865	91·569	−0·079	"	91·490 p2
%s = −4.003%	30·000	11·865	91·691	−0·234	"	91·457 p3
'A' = +6.458%	35·000	16·865	91·814	−0·474	"	91·340 p4
L = 19.376 m	37·511	19·376	91·876	−0·626	"	91·250 p5
RL at T.P.A. = 91.400						

Pro-forma for vertical curve calculation Ch 18·135 to Ch 37·511 Road . A . . .

Fig. 6.17 Example of pro-forma for calculation recording of vertical curves

casionally met on small road projects, the elements of their design is not included within this chapter. The tables previously mentioned should be obtained and kept at head office for reference.

Producing the final kerb lines

Once the road has been pegged and profiled, excavation carried out, and hardcore placed and compacted to a level equal to the underside of concrete kerb race, steel pins of some 10 mm diameter can be used to accurately mark out the back of kerb lines for the kerb layer. From the previously positioned road offset pegs adjacent to the intersection points etc., the engineer can easily establish the back of kerb lines at the corresponding points on the road. This can be done in one of two ways:

(a) by string line, spirit level and tape.
(b) by theodolite, spirit level and tape.

When method (a) is adopted a string line is stretched taut between the offset pegs marking each side of the adjacent chainage or intersection points. The offset distance to the back of kerb line is then taped and plumbed down the excavation with a spirit level and the exact spot marked with the face of a vertical steel pin. (*Note*: for very accurate work and where curve ranging is to be undertaken, it may be preferable to use a wooden peg driven in with its top flush with the hardcore, and a nail used to locate the point.)

With method (b) the same basic procedure as method (a) is used, except that a theodolite is used to produce the line between offset pegs, rather than a string line. Method (b) is useful as it enables the engineer to use the theodolite to check the angle to the opposite peg from a straight, or a particular co-ordinated reference point, and is more accurate especially on roads with a fairly deep cut where plumbing down from a string line would be difficult. Figure 6.18 shown diagrammatically the principles explained in (a) above.

The engineer can proceed to set out kerb lines at chainage locations on curves and straights using the methods explained above, or alternatively use a theodolite and tape to locate intermediate chainage pins between tangent points on curves or ends of

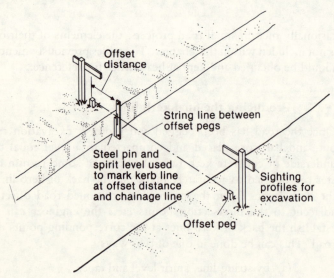

Fig. 6.18   Taping and sighting of offset for kerb line location

straight. The latter method often becomes necessary when chainage offset pegs have become displaced during the course of road excavation. It goes without saying that the engineer should check to ensure that any offset pegs being used to locate final kerb lines have not moved since they were set in position. All critical and important offset pegs should have been well driven into the ground and haunched with concrete to avoid displacement when first established on site.

Even where damage is not obvious, some movement due to ground shift may have occurred. A check on the known measured distance between adjacent offset pegs should give adequate proof of accuracy.

If using a theodolite to set out the kerb line from tangent locations transferred from the offset pegs at each end of a curve and no intermediate offsets remain available, the engineer should set out the curve using the same principles as described for offset pegs i.e., by use of theodolite and tape, sighting along the tangential angle line (see Fig. 6.5). When such a situation arises it will be necessary to calculate the chord lengths for the back of kerb line radius using the formula previously suggested on page 116 and record the results on the pro-forma shown in Fig. 6.4. If accurate records have been kept, as is essential, reference to the calculation sheet used when calculating the same curve's offset pegs will give the tangential angles required. These will of course not alter when points are located on the kerb line and can be copied onto the new calculation sheet direct.

A spirit level should always be used to ensure that steel pins are located in a vertical position and also where taping between pins is on sloping ground. A steel tape marked in 1 mm increments should always be used.

The number of intermediate pins to be positioned will depend on the radius of the curve e.g., a small radius curve of say 15 m or below will need pins at every 5 or even 2.5 m to suffice the kerb layer, whereas on a curve of over 15 m radius, 10 m intervals will be adequate. It is not advisable to locate pins at a greater distance than 10 m, otherwise excessive sag of the string line between points will occur. Obviously the pins will be located whenever possible to coincide with the road chainage and level points.

When set out for line, levels will have to be marked on the pins, and by using the dumpy level, the engineer should mark the pin with the top of kerb level required at each chainage. Levels can either be marked on the pins with pencil, by scribing the bottom of the staff, or readings taken on top of the pin, with the difference in levels then being measured by tape down to the desired points. After marking levels in pencil, which is not easily distinguished on steel, adhesive coloured p.v.c tape should be wound around the pin so that the top of the tape marks the required level height.

Although the kerb layer may now set up his string line to denote line and level of the kerbs, it can be seen from the example in Fig. 6.19 that where curves occur, the stringline runs along the chord line between curve chainage points. When setting up formwork and laying kerbs, the kerb layer should be advised of the measurement for the 'versed sine' of the curve segment in question. The versed sine is the straight distance at the midpoint of the curve between the arc and the chord line (see Fig. 6.1 for diagrammatic information). The value of the versed sine can be calculated by using the formulae given in Table 6.1, or with the use of a programmable calculator by a program similar to the example shown in Fig. 6.20, for use with the Texas TI57 instrument. Given this distance, the kerb layer can ensure that the curve flows smoothly between chainage points.

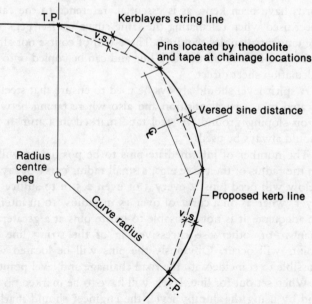

Fig. 6.19 Measuring the 'versed sine' from string line to assist the kerb layer in obtaining a smooth finished kerb line

Program record sheet for Texas TI 57 calculator

Program title:
To CALCULATE THE VERSED SINE OF A CURVE GIVEN VALUES FOR RADIUS AND CHORD LENGTH

Program drawn up by: R.W. MURPHY   Date: OCTOBER 1980

Formulae used: $V.S = R - \sqrt{R^2 - \left(\frac{L}{2}\right)^2}$ where R = radius, L = chord length

Program description and notes

the result is displayed fixed to the nearest millimetre.

Step number	Press	Key code	Calculator display	Step number	Press	Key code	Calculator display
1	LRN	—	00 - 00				
2	RCL 1	33 - 1	01 - 00				
3	÷	45	02 - 00				
4	2	02	03 - 00				
5	=	85	04 - 00				
6	x^2	23	05 - 00				
7	STO 0	32 - 0	06 - 00				
8	RCL 2	33 - 2	07 - 00				
9	x^2	23	08 - 00				
10	—	65	09 - 00				
11	RCL 0	33 - 0	10 - 00				
12	=	85	11 - 00				
13	$\sqrt{x}$	24	12 - 00				
14	STO 3	32 - 3	13 - 00				
15	RCL 2	33 - 2	14 - 00				
16	—	65	15 - 00				
17	RCL 3	33 - 3	16 - 00				
18	=	85	17 - 00				
19	2nd Fix 3	48 - 3	18 - 00				
20	R/S	81	19 - 00				
21	LRN	—	0				
22	RST	71	0				

Information required in memories fed into:
STO 0 – /
STO 1 – Enter chord length value
STO 2 – Enter radius length value
STO 3 – /
STO 4 – /
STO 5 – /
STO 6 – /
STO 7 – /

Fig. 6.20 Program record sheet for calculation of versed sine of a curve

Laying of the road kerbs and concrete base

Adequate setting out will by this stage have been provided to enable kerb laying to commence. However, the engineer should continue to check road lines during this operation i.e., after concreting of the kerb bed and prior to kerb laying.

After kerb laying, the steel pins can be hammered down to below the kerb level and cast in with the concrete backing haunch. Alternatively, if strips of polythene are affixed with elastic bands where the pin lies within the concrete kerb bed, the pins can be pulled up and removed after kerbs are laid and before the backing haunch is applied. Before removing or burying the steel pins, their position should be transferred and painted on to the laid kerbs. These paint marks can then be used by the engineer to assist in the provision of levels for road cambers required at hard core and tarmac stages and will ensure that finish road levels are taken between the correct locations.

To summarize, the engineer should follow the sequence of operations listed below when involved in setting out roadworks.

1. Check all information is available.
2. Calculate locations for setting out points on the centre line.
3. Locate setting out points for the centre line on site.
4. Calculate angles required for offsetting centre line points, and record on file.
5. Offset tangent points and road ends from the centre line.
6. Calculate for intermediate offset peg location on curves, and record on file.
7. Locate intermediate offset pegs at curves and straights.
8. Erect profiles for road excavation.
9. Supervise road excavation and base hardcore.
10. Position steel pins to denote kerb lines on straights and mark kerb levels.
11. Calculate for locations of steel pins to curves, and record on file.
12. Locate steel pins at chainage points on curves and mark kerb levels.
13. Calculate 'versed sine' for use by kerb layer, and record on file.
14. Check pin location after concreting of kerb race.
15. Supervise kerb laying.

16 Remove steel pins prior to concreting of kerb backing, after marking the kerb in paint to show the chainage location.

17 Using painted chainage locations, supervise and check levels of final hardcore and road finish materials.

Note

1 For details of setting out three centre curves, refer to *Kerb Alignment,* by Schneider/Krenz/Osterloh, printed in Germany, 1970.

7

Drainage

Setting out of drainage

Marking the manhole centre-point

The first operation when setting out drainage will be to locate the centre point of each manhole. It is from these centre points that the drain lines are marked out and not from the perimeter of the manhole, it is important to remember this. If drain lines are not set from each centre point, the correct line of channel fittings through the manhole will not be achieved. The information to set out these points may be shown on the drawings as co-ordinate references, marked distances from building or road lines, or alternatively left for the engineer to scale off the plans. The latter method is unsatisfactory, and unless directed to do so by the architect, plans should not be scaled.

Sight rail profiles

Following the installation of centre point pegs at each manhole, sight rail profiles must be erected from which the dig levels can be ascertained. The type of profiles used, i.e., single or double stake, will largely depend on the drain's priority and the sighting distances used. For example, a main drain run more than 15 m in length is better profiled using the double stake and long sight rail method illustrated in Fig. 5.1. On very long drains, intermediate profiles may be necessary. Their levels must be calculated from the plans. Drains of less than 15 m in length and of lesser priority e.g. branch runs from the main to building, can be adequately profiled using a single stake and short sight rail profile as shown in Fig. 6.10 for use on roadworks. The level of the sight rails will be set at the same distance above invert level for manholes at each end of the run to be excavated. This equal height will also be set

146

in relationship to the existing ground levels and contours along the drain's length.

Figure 7.1 shows an example of how ground conditions can dictate the sight rail heights.

Allowance must also be made to permit excavation machinery to pass the profile without disturbance, thus the profiles must be offset an adequate distance from the drain's centre line.

The dig levels of the drain are controlled by using a traveller which is continually checked at points along the drain's length until the top of its sight rail corresponds with the sight line between profiles. The length of the traveller will be constructed to correspond with the distance that sight rails are set above drain inverts at each manhole, with the depth of bed excavation required below the invert level also being added. For example, where drain profile sight rails are set at 2.00 m above invert at each manhole and the drain has a 100 mm bed of gravel, the dig level traveller length will be 2.00 + 0.10 m = 2.10 m. The calculated information regarding sight rail heights and traveller lengths should be entered onto a record sheet as shown in Fig. 3.7.

Checks prior to excavation

After installing the manhole centre pegs and erecting profiles, there are certain checks which should be made prior to final marking and excavation of the drainage lengths.

1. Ensure that the distance between manholes corresponds with that stated on the architect's long sections, plans, or calculations between co-ordinates.

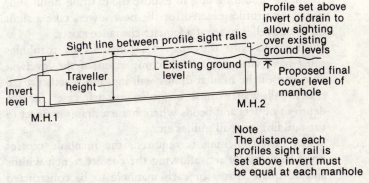

Fig. 7.1 Use of traveller and sighting profiles during drainage excavation

2 On main drainage running under roads or other structures still to be built, programme the route of drainage construction so that such runs are installed as early as possible prior to commencement of these works.

3 As the main drainage is frequently the first item of construction following the reduce dig excavation of the site, check the manhole centre peg locations to make certain they are clear of future works. For example, where manholes are shown alongside a road it may be necessary to temporarily peg the road kerb line to ensure that the manhole, when constructed, does not infringe the roadworks.

4 Check the relationship between exisiting ground level and the top of pipe level at each manhole to ensure that the drain will be of adequate depth to withstand the weight of construction traffic used, prior to the ground being made up to its finished level. Where the drain will be extremely shallow i.e., less than 750 mm to the top of the pipe, it may be necessary to either (a) delay installation of the drain until after heavy site traffic is removed, (b) protect the drain by a layer of concrete above the pipes or (c) lay the drain and fence off each side along its length.

5 Consult the records of existing service routes to ascertain if any cross the drain line. Where these exist, their location must be accurately marked and a trial hole taken to obtain the cable or pipe depth. Overhead cables must be guarded by the erection of safety barriers.

6 Where the new drain may cross or run below an existing sewer, the levels at the point of intersection must be reviewed to make certain that no clash of pipes occurs. A trial hole should be dug to expose the existing drain and, using the profiles erected for the new sewer, take a sight on the traveller to see if adequate clearance exists.

7 Connections into the main via junctions or manholes must have adequate fall. A comparison of the levels at each end of the branch drains will indicate whether a fall exists. Allowance must be given in such cases for the depth of outlets and bends where branch drains connect to the building or road gullies etc.

8 When dual trenching is required, the manhole centres must be carefully set, allowing the correct trench widths and adequate space for each manhole to be constructed without infringement of further drain lengths.

148

Excavation of drainage

The path of the drain running between manhole centres is usually marked by placing a line of white lime on the ground denoting the centre line of excavation. Once the mark has been laid, the manhole centre pegs should be accurately offset to a known dimension in at least one direction. This is best done by placing one peg on each side of the manhole centre peg, so that a line strung between them exactly crosses the manhole centre at a known distance from one or both pegs. Figure 7.2 shows this in diagrammatic form. The advantage of this method is that no instruments are needed to set the pegs at a known angle to the centre peg, and the pegs can be located clear of any obstructions.

Occasionally, when working under difficult conditions, it is not practical to lime out the trench line, in such cases, a string line can

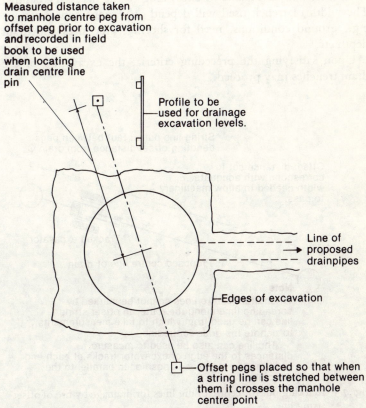

Measured distance taken to manhole centre peg from offset peg prior to excavation and recorded in field book to be used when locating drain centre line pin

Profile to be used for drainage excavation levels.

Line of proposed drainpipes

Edges of excavation

Offset pegs placed so that when a string line is stretched between them it crosses the manhole centre point

Fig. 7.2 Use of two peg offsets for establishment of manhole centres

be pulled between pegs at a known offset distance from the centre line and measurements taken from it to ascertain the actual drain line (see Fig. 7.3).

(*Note*: As a precaution, always check that the excavator used to dig the drain is of the type where the centre of machine is also centre of bucket. On some tracked excavators, this is not always the case as the digging bucket and boom are offset from the machine centre. With such machines, agreement must be made between the engineer and the plant operator whether the machine driver will offset his machine accordingly, or if the lime line should be marked at a calculated offset distance from the manhole centres to correspond with the machine centre line.)

Before excavation commences, the type of machine to be used must be considered in relation to the drainage works to be put in hand. The excavator must be equipped with the necessary range of buckets and be capable of excavating to the necessary depths. The width of trench used will depend on various considerations e.g., ground conditions, need for shoring, pipe sizes and drain depth etc.

Upon satisfying the preceding criteria, the excavation of the drain trenches may proceed.

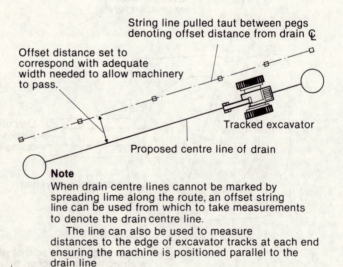

String line pulled taut between pegs denoting offset distance from drain ℄

Offset distance set to correspond with adequate width needed to allow machinery to pass.

Tracked excavator

Proposed centre line of drain

Note
When drain centre lines cannot be marked by spreading lime along the route, an offset string line can be used from which to take measurements to denote the drain centre line.
  The line can also be used to measure distances to the edge of excavator tracks at each end ensuring the machine is positioned parallel to the drain line

Fig. 7.3 Marking out for centre of dig lines for drainage by use of offset string line

Pipe runs and manholes

Following completion of excavation, the invert levels and manhole centre points must be transferred on to steel pins at the base of the dig. Using a string line pulled between the offset pegs at ground level, the distance previously recorded can be measured along the line to locate the manhole centre point. This spot can be transferred to the base of the excavation by suspending a plumb-bob from above or plumbing up a straight edge from dig level. A steel pin of 15 mm diameter should be driven well into the ground to mark the spot. The pin must protrude an adequate distance above excavation level enabling both invert and top of pipe levels to be marked, usually (as with roadworks) by a band of adhesive tape around the pin. Invert levels can be transferred to the pin from the sight rails used for excavation, the distance having been recorded on the drainage record sheet. However, to provide a more accurate level and as a check on the excavation depth, it is preferable to use the level and staff and transfer the level from the nearest T.B.M. On long drains it is necessary to place intermediate level pins.

Before drain laying commences, the pipe bed, usually of gravel or concrete, must be laid to an accurate grade. Frequently, this level is found by the use of 'boning rods'. One rod is held at each manhole invert level with a third rod lined in between these until the top of its cross board corresponds with the line of sight. Laying of the drain pipes may commence as soon as the bed is completed.

Usually (unless laser instruments and targets are used), a string line is pulled taut between the level pins marking the centre line and the level of top of pipe (allowing the thickness required for the dimension of pipe and collar). The first pipe will be laid so that its spigot end is at the correct distance from the manhole centre pin, to allow adequate protrusion inside the manhole shaft. The site engineer should advise the drain layer of the required distance.

Branch junctions

Often juctions have to be laid along the drain's length to accommodate branch drain connections. In some instances, it may be necessary to set junction positions by scaling off the drawing and measuring along the drain from the centre of manhole. Preferably however, markers such as range poles or pegs should be placed at the termination point of the branch drain and the junctions set as near as possible to aim in the direction of these markers. When junction pipes are laid, the angle of the outlet's incline should correspond as near as possible to that required when the final branch drain is laid.

Upon completion of the main drain run's pipework, the location of drain junctions must be measured from the lowest manhole and recorded as explained in Chapter 3. Lengths of timber can also be placed in front of the junction to provide markers during re-excavation when the junction level is almost reached.

Manhole cover levels

Final cover levels for manholes cannot always be accurately ascertained until completion of the area surrounding them. However, the approximate cover level as given on the architect's plans or manhole schedule can be located on the top of a peg alongside the manhole. The level is then transferred by spirit level during the positioning of the manhole cover.

Installation of road gullies

It has been mentioned earlier that checks should be made before the main drains are laid to ensure adequate fall is available from road gully connections. The outlet on a road gully will be at a depth considerably below the finished road level when grating, brickwork, and gully outlet depths are taken into account. The gully is usually located so that it sits inside the kerb lines, the inside edge of the circular pot corresponding approximately with face of kerb line. As the gully may be installed before road kerbs are laid, the engineer will have to mark the proposed face of kerb line adjacent to the gully pot from which the gully is located. It is important that the gully be installed at the correct road chainage point and at the lowest point of each road section in vertical curves etc.

Backdrops

Drain backdrops enable the designer to reduce excavation dig levels, and are frequently used on steeply sloping sites. The backdrop is generally located outside a manhole, and transfers the flow from the higher level of the incoming pipe to the manhole invert. When setting profiles for drainage excavation, care must be taken to ensure the sight rail levels are set to allow for the backdrop at manholes which require this method of construction.

Dual trenching

On main drainage design around modern estates it is common practice to use one split level trench to accommodate both foul and storm sewers. The trench will be excavated at the required width down to the level of the highest pipe first, and then further excavation is carried out over half the trench for the lower pipe. When setting up sight rails for such trenches it is advantageous to use double rail profiles as in Fig. 5.1 with the traveller length being kept constant for both levels of excavation.

Pipe laying lasers

As an alternative to the use of string lines for drain pipe laying, lasers with sighting targets are now widely used. The laser instrument is set at the centre of the lowest manhole and is adjusted so that the line of its laser beam corresponds with the direction and angle of gradient for the drain, and its height corresponds with the calculated distance the laser beam is to be above invert. A target is placed in the pipe at each collar end and the pipe raised or lowered until the visible dot of laser light corresponds with the target centre spot. The instrument offers the possibility of very accurate and fast drain lining.

8

Structures and buildings

This chapter suggests ways in which various types of structures can be set out, and is divided into three main sections:
(A) Traditional structures e.g. brickwork foundations to houses, schools etc.
(B) Steel framed structures.
(C) Concrete framed structures

The main grid setting out and reduce dig will have been completed prior to the actual individual building lines being pegged out on site. The setting out methods explained in this chapter assume this to be the case and cover the stages of setting out from this point.

(A) Traditional structures

As discussed in Chapter 4, on modern constructional plans the corners of buildings are frequently given co-ordinate values and are thus set out from grid stations. When pegs are placed in position to these co-ordinates they mark the actual corner of the building lines, and must therefore be offset clear of the work prior to the foundation excavation commencing.

Profile line boards
The traditional method of offsetting is to transfer the points on to profile line boards set at right angles to each other at each corner of the foundation. The boards are cut long enough to allow the excavation edge lines, centre lines and final brickwork lines to be marked, either by saw cuts or by nails driven into the top of the board. The board is held rigid between two pegs driven well into the ground. An example of profile line board construction is given in Fig. 8.1.

However, the use of profile line boards is both costly in raw materials (i.e. pegs and timber boards) and time consuming to

154

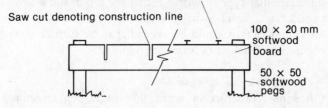

Nails denoting construction lines

Saw cut denoting construction line

100 × 20 mm
softwood
board

50 × 50
softwood
pegs

Elevation

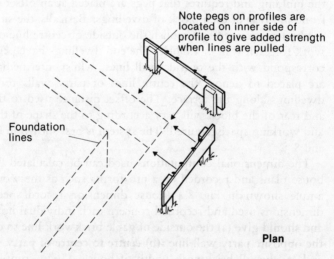

Note pegs on profiles are
located on inner side of
profile to give added strength
when lines are pulled

Foundation
lines

Plan

Fig. 8.1 Line profile details. Elevation of typical single line profile
showing alternative methods of saw cuts and nails to denote
construction lines. Plan of profiles at corner of foundation

construct and set up. To be accurate, the boards must run exactly
at right angles to the lines being marked. Because building design
on many modern sites has clusters of dwellings in groups and
blocks and the line of these blocks is frequently staggered, the
number of boards required can become excessive. The boards
hamper excavation by putting a large obstacle in the path of
machinery. Invariably, when profile line boards are used for such
structures they are knocked out, moved off line or completely de-
molished by excavating plant and dumpers and their reinstatement

can be tedious and time consuming. Their only advantage for this particular type of works is that the top of the board can be set to indicate a level (usually d.p.c. level), as well as construction lines, and thus it could be argued that no further level pegs for concrete or brickwork levels are required.

Single peg technique

On some construction work the single peg technique can be adopted as an alternative and is considered by the author to be far more suitable for traditional construction work. The method can best be explained using terraced housing construction as an example.

This method rules out the need to place profiles at all corners of the building and requires that pegs are placed at an offset distance front and rear of each block of dwellings. Basically the single peg method is arranged to establish the outside or centre line of brickwork for each wall in a block. The end dwellings have pegs set to correspond with the outside wall lines, whilst intermediate pegs are placed to denote the centre lines of party walls to internal dwellings along the terrace. The offset distance used to the front and rear of the block will vary according to the shape of the block and working space required. The system is best shown in Figs. 8.2 and 8.3.

The dimensional computations used can be calculated from the house plans and recorded on a pro-forma such as the worked example shown in Fig. 8.4 (house dimension record sheet). The dimensions used and recorded concern each individual house type and should give (a) the outside of gable brickwork line to centre of the opposite party wall line, (b) centre to centre of party wall line and (c) overall brickwork width of house. Upon completion of the house dimension record sheet, the engineer can prepare his setting out sketch for each block, indicating the distances to be used during setting of the pegs as in Figs. 8.2 and 8.3.

The single peg technique has the following advantages over the profile line board method:

1 a saving of raw materials e.g., profile board timber, nails, pegs etc.
2 a considerable saving in time
3 produces a standard procedure which is easy to remember, record and follow
4 permits straight line and right angle setting out

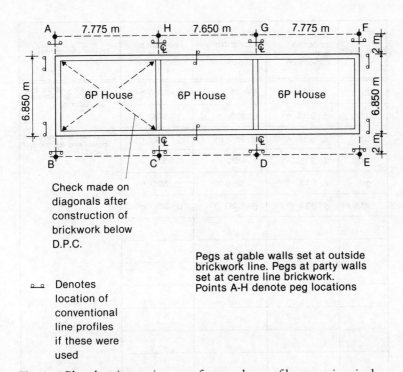

Check made on
diagonals after
construction of
brickwork below
D.P.C.

Pegs at gable walls set at outside
brickwork line. Pegs at party walls
set at centre line brickwork.
Points A-H denote peg locations

⌐⌐ Denotes
location of
conventional
line profiles
if these were
used

Fig. 8.2 Plan showing setting out of terraced row of houses using single
peg technique

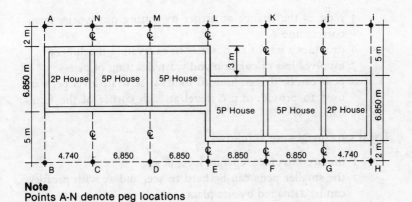

Note
Points A-N denote peg locations

Fig. 8.3 Plan showing setting out of a staggered terraced row of houses
using the same principles as detailed on Fig. 8.2

House type	Outside wall to centre wall dimension	Centre wall to centre wall dimension	Overall outside to outside dimension	Notes and remarks
2P	4.740 m	4.615 m	6.850 m	1. 250 mm gable and party wall widths required
4P	5.925 m	5.800 m	7.300 m	
5P	6.975 m	6.850 m	6.850 m	2. Centre line remains continuous at stagger condition
6P	7.775 m	7.650 m	6.850 m	
8P	8.685 m	8.560 m	7.800 m	
3P FLAT	5.925 m	5.800 m	6.850 m	

Fig. 8.4 Worked pro-forma for house dimension record sheet

5 reduces the chance of errors by setting out works to incorrect lines
6 provides a less of an obstacle to excavating machinery
7 involves less calculation and wasted setting of pegs
8 pegs can be driven in until solid, as separate level pegs are used to provide d.p.c. level at each corner of the foundation.

The disadvantages are as follows:

1 the need to erect separate level pegs
2 the smaller pegs can be hard to see, and as with profiles, can be damaged by site plant
3 necessitates long offset dimensions when used on staggered blocks.

Generally, it can be seen that the advantages outweight the disadvantages and therefore prove the system's merits.

Marking and levelling the foundation excavation and brickwork lines from single peg setting out

Excavation line and level

When setting out pegs have been positioned, the excavation lines can be marked and foundation dig may commence. The centre lines of foundation are usually marked, unless the trench is wider than the machine's bucket widths, in which case the edge lines are denoted. When marking the centre lines, string lines are pulled between the opposite pegs and a thin line of white lime is placed on the ground below the line. This applies to intermediate pegs only and the gable ends must have steel pins offset from the pegs (which denote face brickwork lines), between which the line is pulled. The front and rear of each dwelling will also be marked from a line strung between pins placed by measurement from the offset pegs. When edge of foundation lines are marked, steel pins are measured each side of the centre line or face line pegs and both sides of the trench limed out on the ground.

The level for excavation can be given by one of two methods. Method (a) involves the use of sight rail profiles, such as the type used for drainage works together with a traveller to denote dig levels. Method (b) can be used if an automatic level is made available for use by the plant operator and banksman.

At the commencement of excavation for each row of houses, the engineer sets up the level to a known height of collimation. As the level will need no further adjustment after its initial setting, the levelling staff can be marked by an elastic band or p.v.c. tape to denote the dig level, or alternatively a length of wood can be marked at the height required. The level is left for the plant operator and banksman to use during excavation of the foundation, which continues until the dig level corresponds with the staff, or the rod mark equals the height of collimation.

Foundation concrete levels

As each section of the excavation is completed, the foundation concrete levels will have to be established. The levels must be set accurately from the nearest T.B.M., and need to be on a solid marker, driven well into the base of the trench. Perhaps the best method for concrete level markers where concrete is less than 750 mm deep, is the use of removable rods of 20 mm diameter steel bar, with cross pieces welded to the upright in two places. The lower cross piece denotes the concrete level, whilst the upper piece enables the pin to be removed after pouring of the concrete, but prior to its initial set. An example of this type of level pin is shown in Fig. 8.5.

Where concrete is deeper than 750 mm, shorter steel rods can be driven into the sides of the trench to mark the top of concrete level – these usually being left in after the concrete is cast.

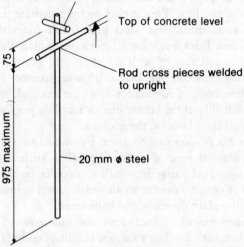

Higher cross piece used as grip to pull out pin whilst concrete is still wet.

Top of concrete level

Rod cross pieces welded to upright

75

975 maximum

20 mm ∅ steel

Fig. 8.5 Example of steel rod used to denote concrete levels in foundations

Brickwork line and level

After the concreting of the foundation, the corners of brickwork are set out by line and tape. Using a theodolite or string line, a line is taken between opposite pegs, and together with the taped offset distance, the corner or edge line of the building is marked. The mark on the actual foundation can be made by spreading a thin layer of mortar over the area and marking in it a line using a trowel (see Fig. 8.6). If the top of the foundation is at a lower level than the ground, the marked position must be accurately

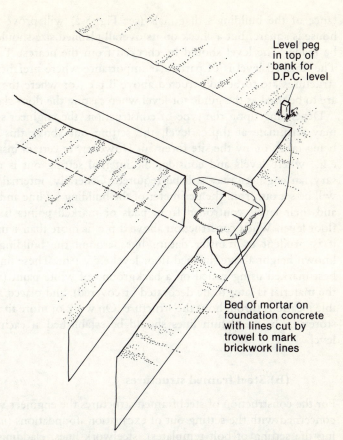

Level peg
in top of
bank for
D.P.C. level

Bed of mortar on
foundation concrete
with lines cut by
trowel to mark
brickwork lines

Fig. 8.6 Section of foundation trench showing method of marking
brickwork lines in a thin mortar bed

plumbed down from the line and tape using a spirit level. Care
must be taken to ensure that the correct side of the level is used
when lining in and marking.

The level for brickwork coursing is usually given by a peg
driven into the ground alongside each corner of the foundation.
Either the peg top, or a nail driven into its side can be used to denote the level. Frequently, the damp proof course level is given,
from which the bricklayer uses his spirit level to transfer the mark
to the actual corner of the building, and to establish the brickwork
gauge.

Upon completion of the brickwork, a check should be made to
ensure that the size and level of the walls are accurate. A taped dis-

tance of the building's diagonals (see Fig. 8.2) will prove if the house is square, but a check on its overall planned size should also be made. The level should be checked from the nearest T.B.M. The finished level is of particular importance where prefabricated structures are to be constructed above d.p.c., or where the walls are to be used as the guide for level when casting the floor slab.

Depending upon the type of construction, the engineer's tasks may terminate at d.p.c. level, the setting out above this point being taken on by the site foreman. However, in certain instances e.g., where levels and considerable internal setting out is necessary, an engineer may still be required. Generally, internal walls will be set out from the perimeter of the building by line and tape, and their levels controlled from pads or marked points fixed at floor level. When construction above d.p.c. is more than 1 m high it is prudent to transfer datum lines around the building at a known height above finished floor level e.g., 1 m. These lines are best marked using pencil on a background of white paint (where the material is later to be decorated or covered), and placed at visible locations throughout the structure. On work of more than one storey high, the datum lines should be established at each floor level.

(B) Steel framed structures

For the construction of steel framed structures the engineer will be concerned with the setting out of excavation, foundations (including the setting of bolt templates), steelwork lines, cladding lines and levels. Because pre-fabricated components have little or no tolerance, precise location of holding down bolts or pockets, in both horizontal and vertical alignment, is essential. Although it is usual to cast the bolts in a sleeve to allow a degree of flexibility, accurate location is still necessary. The sleeve's flexibility provides useful tolerance for steel erection and final lining purposes.

Obviously with the use of pre-fabricated components, any errors found during the course of construction will be exaggerated over the whole length of the structure. The longer the building the more critical becomes the need to keep a close watch on the location of components. It should therefore be appreciated that errors occuring by taping from say, base 1 to base 3 for the stanchion bases in Fig. 8.7, and then base 3 to base 5 and so on, could form a cumulative total error if inaccuracies occurred in each measured length, and this method should be avoided. Wherever

162

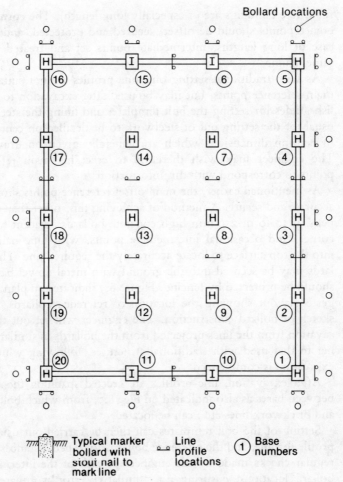

Bollard locations

Typical marker bollard with stout nail to mark line — Line profile locations — (1) Base numbers

Fig. 8.7 Typical stanchion base layout for steel framed structure indicating location of setting out bollards and line profiles

possible, the structure should be set out on a running tape distance. The same problem can, of course, occur if discrepancies are evident in the prefabricated components used. If, in the case of connecting beams between columns, each beam was 3 mm short or long, then a cumulative error of 30 mm would occur over the space of 10 bays.

With these problems in mind, the engineer must set out the structure as a whole from end to end and side to side, and ensure that the overall lengths are as required (E.D.M. equipment is useful where buildings are of especially long length). The corner reference points should be offset, secured and protected, and in the case of long lengths, intermediate points set and treated in the same manner.

As with traditional setting out, line profiles are not suitable as main reference points, but may be used after excavation to establish guides for setting the bolt templates and lining the steel. It is usual for the setting out of steelwork to be detailed on centre line of column dimensions which are generally given on drawings. The engineer may wish therefore to erect his main reference points to correspond with this information.

As mentioned earlier, the main offset reference points should be accurate and secure. A method of achieving this, other than secure pegs, is to form 250 mm high concrete bollards adjacent to each corner, and occasional intermediate points, with pipe nails cast into the top surface to locate accurately the required line. The bollards may be secured into the ground with metal dowel bars and should be protected by fencing. Figure 8.7 indicates on plan a suggested layout showing the location of reference bollards, and a sketch of bollard construction. The engineer will set out the excavation from the lines projected from the bollards in similar manner to that used with traditional structures, following which the excavation is commenced.

After excavation, line profiles are erected spanning the trench between bases as also indicated on Fig. 8.7, from which bolt, steel and brickwork lines etc., can be located.

Setting of the bolt templates can then be carried out using the profile for line, and the reference bollards for taped distances, with regular checks made by the engineer at each of the intermediate bollard locations, ensuring no cumulative error is apparent. A typical bolt template is shown in Fig. 8.8. Each template must be levelled, and this can be achieved by initially placing a level datum on one of the template support pegs, which is later transferred by the carpenter to the centre of template point by using a spirit level. The engineer must remember to make adequate allowance on the concrete level for any packing shims required below the steelwork baseplate, and check the drawings thoroughly to ensure such allowance has been included in the design levels. Concrete levels for the trenches between bases can be established by the steel pin method described for traditional structures.

Having erected the formwork to the bases, the concrete can be placed, but during pouring, and again before the concrete has set,

164

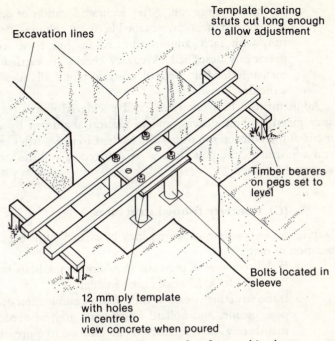

Fig. 8.8 Typical detail of bolt support template for stanchion base construction

the engineer should check the line and level of each template and profile to ensure. that no disturbance has taken place during casting, and he may supervise any adjustment required.

When the actual steel frame is to be erected, the usual involvement for the engineer is as follows:

1 Steel level plates known as 'shims' have to be placed and levelled on each stanchion base to the required underside of column level.

2 The columns and beams are erected and bolted into place to the approximate line given by stringing between the previously erected profiles, together with the taped distances on a running tape from the offset bollards.

3 After fabrication of the overall structure, accurate lining and plumbing of the works must take place. The engineer should erect a theodolite at a known distance from the edge of each column line e.g., 50 mm, and line in each column for vertical and horizontal exactness using a tim-

ber offset gauge rod. After accurate location of one corner, the distances can be taped between columns along the run, with each column adjusted for line to its edge as previously described. Accuracy checks are carried out at each intermediate bollard position along all sides of the structure.

Adjustment in horizontal line can be achieved within the tolerance given by bolt sleeves, and in vertical alignment by adjustment of the bolts on opposite sides of the baseplate. After erection and adjustment, the bolt sleeves are grouted in and the structure accurately secured.

(C) Concrete framed structures

Setting out of concrete structures may be carried out using a combination of methods:

1 If the structure is pre-cast, then similar methods to those described for steelwork may be undertaken.
2 If the structure is cast in situ, then a combination of single peg, profile, and bollard setting out may be needed, remembering that reinforcement location requires careful consideration.

For the average size builder, concrete structures require a higher degree of control than any other work. The structures usually contain little or no tolerance, and it is vital that the correct location of reinforcement and finished work is achieved. The engineer will be responsible for providing adequate setting out information to ensure these correct positions, and must therefore take sufficient time for the preparation of details, setting out sketches and peg location on site, to ensure that work is cast accurately.

Remedial work to overcome errors in this type of construction can involve a great deal of expense and lost production time.

Setting out for excavation will be exactly as explained for traditional work and steel frame work, as will control of level and formwork lines.

After excavation and location of reinforcement, but prior to concreting, the engineer should ensure that the work has been carried out to the correct levels and positions, also that correct reinforcement has been used and has adequate cover. As with steel framed structures, building lines must be checked before placed concrete has set. This is particularly important when formwork is used, as its line can be displaced during the pouring of concrete.

The engineer will be involved in the levelling of floor slabs, column heights and retaining walls etc., which will often be carried out at various locations on the contract. Concrete heights inside formwork can be marked by a pencil line on the face of the formwork or by driving nails through the timber to denote the level. As many secure reference points as is possible must be established around and within the construction area for both lines and levels.

Often concrete framed structures are dimensioned on a grid system, and it is advisable to locate reference points at positions on that grid, or alternatively at corresponding offset locations to it. The points chosen should be those unlikely to be disturbed during the course of the work. Failure to provide adequate and permanent reference points will only result in defective work, and require costly remedial measures, and the engineer will have considerable difficulty in achieving dimensional accuracy.

Vertical alignment on tall multi-storey work can be controlled by the use of an autoplumb if desired.

9

External services and external works

Services

Service main installations on average projects generally involve the laying of water, gas, electricity and telephone supplies. Sometimes other services such as T.V. and fire alarm systems etc., may also be specified. Depending on the sequence of installation, the engineer may have to set out for each individual service, or for one common trench in which all supplies will be accommodated.

Time to install

The period during which services are laid will vary with the type of project in hand. On a factory or school site, services are frequently left until the external scaffolding is dropped. Conversely, on a housing site the deepest service mains such as water and gas can often be laid after the oversites are cast and before superstructures commence. Electricity, telephone mains and all branch services to each house would then be laid just prior or during the construction of external works.

In general the installation period will have to be determined at the planning stage relative to the proposed construction.

Sub station/governor houses

On large projects where the electricity and gas installations may be split into sectional areas, it is often found that sub stations, transformer enclosures or gas governor houses are specified in the design. The engineer will invariably have to set these out at a fairly early stage of the works to allow installation of the services equipment prior to sections of the installations being made live. Details of their construction and location will be provided either by the architect or the statutory authority concerned.

Organizing the services

Usually the services will be supplied and laid by each statutory authority, and it is essential that they are organized by the contractor, who must endeavour to have each service laid consecutively. It is advantageous to all parties if a services meeting is held on site prior to installations commencing. A representative of each statutory authority should be invited to attend, and the site engineer should be present along with the contractor's other senior staff. Dates for installation of each service should be given to the various authority representatives and on large sites this may be broken down into areas.

Frequently the service routes may have to differ from those originally proposed by the architect and therefore detailed plans should be requested from each service authority showing their official routes. The location of existing services must be obtained and the statutory authority may offer to trace these on site if their actual location is somewhat dubious.

The contractor's representative who chairs the meeting should ensure that minutes are taken and distributed to all relevant personnel.

Setting out for services

Before the services are installed the existing ground should be excavated to its approximate reduced level, allowing the correct depth of dig to be excavated by the installers. As services frequently run alongside roads or in footpaths, the setting out for reduced dig will involve sight rails, erected at each change of level in the path, and the marking of path edge lines. Depending on the type of installation proposed i.e., single or common trench system, each service may have to have its trench centre line marked, or alternatively the centre line of the common trench may be marked. The level for excavation can be taken from the sight rail profiles erected when the footpath was reduced, although the traveller length should be altered to give the correct dig levels for each service. It is important that each service is laid at its correct location and level within the trench area.

On housing work, branch services will feed each house from the main. It is helpful if the service trench for these is dug along with the main trench, or alternatively left until after main installations if each authority has chosen to excavate and lay individual services. Some final connections to the mains i.e., water, street lighting and telephone etc., are frequently left until the footpath edgings are laid. From these, exact heights can be obtained for stop tap covers, joint posts or street light column base levels. However, the branch pipe for the water service, is better laid prior to the edgings, leaving only the connection joint to the main to be completed. Likewise, if the telephone services are ducted from the main to each house, these ducts should also be laid first.

The engineer should remember that heavy cables are difficult to bend around tight corners, and it may be necessary to allow for this when setting out. Care is needed in such circumstances to ensure that the service does not move out of the designated service zone e.g., the width of the footpath.

Plotting after laying

Immediately following the installation of each service, its position should be determined relative to the nearest stable point of known location. The information should then be recorded on a site plan which can be officially issued as a record drawing by the architect upon completion of the service installations.

Where services pass under roads, the cables are usually laid in pipe ducts, however, water and gas mains may alternatively have lengths of service pipe installed rather than a duct. Whichever method is used, the ducts or pipes must be positioned prior to the road tarmac being laid, but preferably after setting of the kerb race or kerbs and after the initial hardcore base has been placed. The location of these should be marked on the layout plans and also by using coloured paint on the kerb directly above the duct. A different colour should be used to represent each service, e.g.:
   (a) blue for water
   (b) red for electricity
   (c) green for gas
   (d) orange for street lighting
   (e) white for telephones.

Other colours should be used for each additional service, and the depth of each duct or pipe should also be noted.

External works

The amount and type of external works will vary with each contract, but can include setting out for the following items:

    (a) screen and retaining walls
    (b) footpaths (tarmac, concrete, brick, hoggin or paving slab)
    (c) car parking bays
    (d) sheds and other sundry structures
    (e) fences
    (f) play areas
    (g) landscaping
    (h) earth mounding.

The setting out procedures used for each of the above will only be explained briefly as the general principles are similar to those discussed in previous sections of the book.

(a) Screen and retaining walls

The location of screen or retaining walls (a retaining wall being one which supports material at a higher level at one side than the other), are often planned to line in with sections of the main building. Alternatively, the architect may give distances from the building lines to the walls. Where no dimensions are given their location may have to be scaled from the plans.

The concrete foundations and brickwork are set out in the same manner as the method described for the main building substructures, and the brickwork levels can be transferred from the main structure to ensure that brick courses run parallel.

(b) Footpaths

As with external walls, the location of footpaths may or may not be dimensioned on the plans and frequently can be adequately set by scaling from the drawings. Where service installations run in the paths, this scaled distance should correspond with that used at the time of service installations. The path lines can be roughly set for reduced dig purposes and sight rail profiles used with a traveller to establish excavation levels. The actual path edging kerb, sett or timber board can be denoted by steel pins between which a string line is pulled to levels marked by adhesive tape. The procedure is generally identical to that used when setting out for road kerbs.

It is important that adequate and correct falls are given to the footpaths. Once the edging has been laid and is solid and upon completion of any remaining service installations, the hardcore base can be laid ready to receive the footpath topping. A scratch board can be used to establish the correct hardcore level such as the types shown in Fig. 9.1. It should always be remembered that footpath lines often have to correspond with building and vision splay lines, and should therefore be carefully set out.

(c) Car parking bays

Depending on their construction, some car parking bays may have been set out and constructed at the same time as the main road-works, but these can be left until the external works period of the project. If the bays are constructed early, then damage may result during the course of the works but the bays may be used as ma-

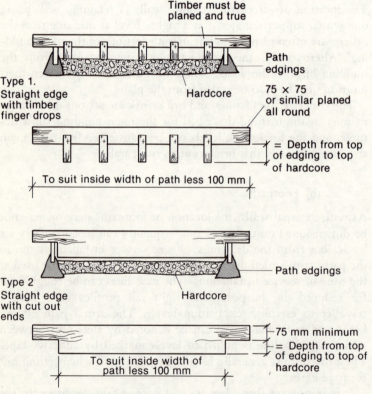

Fig. 9.1 Alternative methods of scratch board construction for use on footpaths

terial storage areas. Conversely, if constructed later then adjustment may be required to the drop kerbs forming the channel line of the main road, but the risk of damage after construction is drastically reduced. The engineer will have to evaluate the best procedure for the site in question.

The procedure for setting out the kerb lines will be as described for roadworks and footpaths.

(d) Sheds and other sundry structures

The methods used for setting out sheds and other ancillary structures will be as explained for the setting out of the main structures. Sheds, however, may have to be adjusted slightly to correspond with completed superstructure lines of the new buildings.

(e) Fences

Fence lines and post holes can be set out in accordance with their planned location, ensuring that where a sight line corresponds with that of the building, the correct line is taken. Often the only way of establishing the fence lines is by scaling off the plans, and this should generally have the architect's approval prior to the setting of pegs on site. Usually the topsoil will be roughly laid prior to the fencing work being completed and the ground will therefore have been reduced to the correct levels, and the fence height will normally follow the ground level.

(f) Play areas

Play areas should be set out at the same time as footpaths and completed with each sectional or single handover area. On some sites it is prudent to leave the installation of the equipment until each area is completed.

(g) Landscaping

More often than not, the contractor is responsible for spreading and levelling the topsoil. The topsoil level should be controlled using sight rails and travellers. The engineer must make allowance for the depth of topsoil required, which generally varies between grass and shrub bed areas.

(h) Earth mounding

Landscape schemes may include earth mounding. Guides to the levels required can be established using stakes, either hammered in so that their top corresponds with the finished level, or painted up to the level required. The stakes can be knocked in at various levels of the mound until the finished height is reached. Batter rail profiles should be used if mounds are of considerable height and require graded sloped sides.

Glossary

Terms and abbreviations relative to site engineering

Alidade
> Plate supporting the trunnion standards of a theodolite, incorporating the plate level.

Apex
> During curve ranging, denotes the point of intersection of the curve's tangent lines, being at the peak of the triangle formed (see Fig. 6.1).

Arc
> Part of the circumference of a circle or other curve.

Arc length
> The length of arc between tangent points.

Backsight (B.S.)
> When levelling, the first sight taken after setting up the instrument.

Baseline
> An accurate line used as the starting length for a survey or setting out.

Bearing
> The angular orientation of an object from either:
> (1) True North, (2) Magnetic North, (3) Grid North
> (4) Arbitrary North, Meridians.

Bench mark (B.M.)
> A point of stable base and known level used as a datum for levelling. Ordnance Bench Marks (O.B.M.) are located throughout the country and often marked by the B.M. sign

Boning rods
> Set of three T-shaped timber rods used to set out an even grade

of sight between two fixed points. Two rods are held on the point of level at each end whilst the third rod is lined in to correspond with the line of sight at any point between them.

British Standard (B.S.)
A set of publications by the British Standards Institution which describe the standards and methods of testing materials etc.

B.S.C.P.
British Standard Code of Practice.

Building Research Establishment (B.R.E.)
The establishment frequently issues digests concerning the construction industry.

Camber
Amount of rise on a convex upper surface, often at the centre of a road or precast concrete beam.

Change point (C.P.)
Used during levelling as a point of temporary datum whilst the instrument is moved to a new location. Two readings are thus taken on the same point.

Catenary
The sag curve into which a tape falls due to its own weight when pulled between two points.

Centering rod
A feature on some makes of theodolite tripods, used when centering the instrument over the set up point.

Chainage
A length measured by chain or steel tape, often used in describing the distances along a road centre line.

Channel line
Term used to denote the edges of a road running alongside the kerbs.

Chord
The straight line connecting two ends of an arc.

Contour
A line drawn on a map or plan to connect all points of the same level.

Cover level (C.L.)
The level given on drainage drawings being the finished height of manhole covers.

Crossfall
The amount of slope across a width of road or other sloping flat surface.

Datum
> A point of reference of known level which may be arbitrary or taken from an Ordnance Bench Mark (O.B.M.).

Deflection angle
> Angle between the lines extending from the radius point to the two tangent points of a curve (see Fig. 6.1).

D.M.S. (Degrees, minutes, seconds).
> Button used on Texas TI57 calculator to convert decimal degrees into degrees, minutes and seconds or vice versa.

Diagonal eyepiece
> Attachment fitted to an instrument's telescope eyepiece primarily for viewing steep sights and plumbing.

Dip
> Term used when measuring below a string line pulled between two points, generally undertaken when checking hardcore or tarmac levels.

Electromagnetic Distance Measurement (E.D.M.)
> Electronic distance measuring instruments frequently termed 'distomats' used to measure distances between two points when surveying or setting out.

Face left (F.L.)
> The position of a theodolite viewed from the eyepiece side when the vertical circle is to the left of the telescope.

Face right (F.R.)
> The position of a theodolite viewed from the eyepiece side when the vertical circle is to the right of the telescope.

Finished floor level (F.F.L.)
> The final floor level of the building usually given a level value.

Foresight (F.S.)
> When levelling, the last sight taken before moving the instrument to a new location.

GTO (Go to)
> Button used on Texas TI57 calculator to instruct the instrument to move to a particular labelled section of a program i.e., to a particular point of a program.

Ground level (G.L.)
> The finished level of existing or proposed ground levels.

Horizontal curve
> A curve on plan, where all measurements are considered in the horizontal plane.

Intermediate sight (I.S.)
> The term given when levelling to any readings other than the first (backsight) or last (foresight).

Intersection point (I.P.)
> The joint location of two straight lines of different bearings.

INV (Inverse)
> Button used on Texas TI57 calculator to reverse the function of a button e.g., a displayed tangent value of a number can be converted back to the original number by finding the inverse tangent of the displayed figure.

Invar band
> A tape made of invar metal

Invert
> The lowest finished surface of a drain, although sometimes taken as the channel level at the centre of the manhole.

LBL (Label)
> Button used on Texas TI57 calculator to label by application of a number, any point in a program, if desired.

LRN (Learn)
> Button used on Texas TI57 calculator to put the instrument into 'learn' mode prior to inserting the program steps into the calculator, and to take the calculator out of learn mode when the program has been inserted.

Optical plummet
> Eyepiece provided on some theodolites through which the set up point directly below the centre of the instrument can be viewed.

Ordnance Survey (O.S.)
> The government agency responsible for mapping in Britain.

Parallax
> The apparent difference in an object's position or direction detected by sighting at different angles through an instrument eyepiece.

Plumb-bob
> A pointed weight supported on a cord suspended from above to mark the point vertically below.

Profile
> Sight rails supported on stakes set to a known level above excavation or fill areas. Used with a traveller which is lined in with the sight lines between profiles. Alternatively a board used when marking building lines for construction work.

Quadrant bearing (Q.B.)
> Angles measured to the east or west of the north/south meridian (see page 86).

Radius (*R*)
> The distance from the centre to the perimeter of a circle.

RCL (Recall)
> Button used on Texas TI57 calculator to recall a value from a memory store.

Reduced level (R.L.)
> An elevation calculated from an agreed datum.

Reference point (R.P.)
> The point used as the commencing line for a theodolite survey. Usually the most distant point of known position.

RST (Reset)
> Button used on Texas TI57 calculator to reset a program at the first step.

R/S (Run/Stop)
> Button used on Texas TI57 calculator to start and stop a program fed into the machine.

Scale factor (S.F.)
> Correction applied to convert projection distances to ground distances.

Sight rail
> Board fixed horizontally to a profile stake to mark a height of given value. Used to denote the sight line between two profiles from which, by the use of a traveller, the finished levels will be ascertained.

Stadia lines (or stadia hairs)
> Two horizontal lines above and below the central line of sight through a level or theodolite telescope used to obtain staff readings for distance calculation (see also Tacheometry).

Station
> A reference point of known location e.g. a grid point of given co-ordinates.

STO (Store)
> Button used on Texas TI57 calculator to enter displayed numbers into the various calculator memories.

Tacheometry
> The operation of measuring distance with the stadia hairs of an instrument, usually a theodolite, using a measuring staff.

Tangential angle

Used during curve ranging to denote the angle between the tangent line leading to the apex of the curve, and a chord on the curve.

Tangent point (T.P.)

The position at which a curve becomes a straight or changes its curvature.

Temporary Bench Mark (T.B.M.)

A solid datum or number of points often transferred on to a site from the nearest B.M. and used as a reference base for the levels on a project.

Traveller

A T-shaped staff of measured length used to give the depth of excavation when the top of its cross rail corresponds with the sight line between profile sight rails.

Tribrach

The triangular frame supporting a theodolite mounted on three adjustable levelling footscrews.

Trivet stage

The mounting plate for the three levelling footscrews supporting the tribrach of a theodolite.

Trunnion (transit axis)

The horizontal axis mounting of a theodolite telescope known as the transit axis held at each end in a fork called the trunnion.

Versed sine

The distance from the mid point of a chord to the edge of a curve defined by the chord's length.

Vertical curve (V.C.)

Used during roadworks to provide a gradual change in a vertical plane between two intersection lengths of a road of different slopes or gradients.

Whole circle bearing (W.C.B.)

Angles measured clockwise through 360° from the north.

General bibliography

BS 5606: 1978 *Code of Practice for Accuracy in Building,* British
Standards Institution, 2 Park Street, London WIA 2BS.

BRE Digest 202 (June 1977) *Site use of the theodolite and surveyors
level,* Building Research Station, Garston, Watford WD2 7JR.

C.D. Burnside, *Electro-Magnetic Distance Measurement,* Crosby,
Lockwood, Staples, 1971.

CIRIA Manual of Setting Out Procedures, Construction Industry
Research and Information Association, 1976.

Schneider/Krenz/Osterloh, *Kerb Alignment,* Bauverlang GmbH,
Wiesbaden (Berlin Library of Congress Catalogue Card No.
72-75933).

F.A. Shepherd, *Engineering Surveying Problems and Solutions,*
Edward Arnold Publishers Limited, 1977.

J. Uren and W.F. Price, *Surveying for Engineers,* Macmillan,
1978.

Index

183